MEASURING THE QUALITY OF EDUCATION

A Report On Assessing Educational Progress

Willard Wirtz
Archie Lapointe

MEASURING THE QUALITY OF EDUCATION

A Report
On Assessing
Educational
Progress

MEASURING THE QUALITY OF EDUCATION

A Report On Assessing Educational Progress

Willard Wirtz
Archie Lapointe

Wirtz And Lapointe
Washington, D.C.

This report was made possible by grants from the Carnegie Corporation, the Ford Foundation and the Spencer Foundation. It does not, however, necessarily reflect the views of those agencies.

Copyright ©1982 by Wirtz and Lapointe.

All rights reserved. Printed in the United States of America.

Design By Mary Kaye Nilan

 300

CONTENTS

FOREWORD	ix	PART V	53
INTRODUCTION	xiii	Implementation	
PART I	1	CONCLUSION AND SUMMARY	59
The Assessment And Educational Standards		APPENDIX A	65
PART II	13	Examining NAEP's History, Goals, Methods, And Previous Evaluations	
Perfecting The Assessment			
PART III	39	APPENDIX B	75
Assessment, Analysis, And Policy Development		We Owe Thanks To...	
PART IV	47	REFERENCES	79
The Overview Function			

FOREWORD

FOREWORD

Although this report will inevitably be taken as passing judgment on the National Assessment of Educational Progress (NAEP), this will miss both its purpose and any real value it may have. What is set out here will be worthwhile only as it informs the political process that is shaping the Assessment's future course. Realizing this has prompted particular effort to make the report more than an expression of merely personal points of view.

The report reflects most significantly the carefully considered conclusions of the Council of Seven which was established at the beginning of the project. Selected primarily for their recognized responsibility and good sense, they also reflect a variety of experiences and institutional interests: Gregory Anrig, then Massachusetts Commissioner of Education, and now President of the Educational Testing Service; Stephen K. Bailey, who is the Francis Keppel Professor in Educational Policy and Administration, Harvard Graduate School of Education; Charles Bowen, Director of Plans and Program Administration for University Relations, IBM Corporation; Clare Burstall, Deputy Director, National Foundation for Educational Research in England and Wales; Elton Jolly, Executive Director, Opportunities Industrialization Centers; Lauren Resnick, Codirector, Learning Research and Development Center, University of Pittsburgh; and Dorothy Shields, Director of Education, AFL-CIO. At least six Council members participated in each of the meetings which were convened: at the outset, during the course of the project, and to review and reshape the final report.

These meetings with the Council were extraordinary personal and intellectual experiences. The inquiry and discussion started from examination of the broad concept of educational assessment, and this became the context for considering the National Assessment itself. It was the Council's suggestion and eventually its decision to shape the entire report in terms of the Assessment's potential role in developing higher and more effective educational standards. Where we had been timid about this, the Council moved boldly. They were right.

None of the individual Council members would have said all that is in the report precisely as it appears, or have omitted all that has been left out. The report represents, however, what became to a remarkable extent the common judgment of these seven outstanding individuals.

Special analyses of particularly important points were requested from five other individuals with special competencies. Edward C. Bryant wrote about the measurement of change in national achievement, and Carol Gibson about the relationship of a national assessment to the improvement of educational opportunity for all students. Patricia A. Graham considered the subject of changes in the educational environment of the United States from 1964-1981. Frederic M. Lord looked at the advantages and disadvantages of the current

FOREWORD

assessment design. P. Michael Timpane analyzed the feasibility of establishing a national consensus on what students should learn. These papers influenced the report substantially.

When the project was about two-thirds along, a working memorandum suggesting tentative conclusions and recommendations was circulated among some 130 individuals whose diversity of opinion assured a spectrum of reaction. Seventy responded, in almost every case carefully and thoughtfully. These replies are invaluable contributions to the political process involved here and are being preserved in their original form to permit further use in this connection.

The radical differences between that working memorandum and this final report testify to the calibre of the responses to it and the seriousness with which they have been taken. The dominant reaction was that educational standards are here to stay, and that the Assessment's future depends on its capacity to improve their effectiveness. The emergence of this consensus in these responses permitted proceeding with confidence to the conclusions reported here.

Appendix B lists those with whom this subject was taken up in one form or another. Special advantage was taken of the generosity of time and mind of Lee J. Cronbach, Stanford University; Frederic A. Mosher, Carnegie Corporation; P. Michael Timpane, now Dean of Teachers College, Columbia University; and David E. Wiley, Dean of the School of Education, Northwestern University.

To add an express assumption of full responsibility for what have to be in the end personal value judgments is largely amenity. Our personal views differ in no material respect from those which evolved in the Council deliberations and from most of the responses to the working memorandum. What is important is that there is a commonality of informed and constructive judgment providing firm basis for political decision.

Serious regard for that process has prompted being as clear as possible about the uses and the users not only of the Assessment but of this report. Some of the recommendations made here would permit direct response by the NAEP Assessment Policy Committee (APC), which governs the Assessment. The report has been prepared in the light of extensive discussion with APC and Education Commission of the States (ECS) representatives and with the prospective exercise of their responsibility in mind.

Under present Congressional authorization, the officers of the National Institute of Education (NIE), in the federal Department of Education, can influence the direction in which the Assessment moves. Some of the criticism of the Assessment has come from the NIE. The report is intended to be directly responsive to these criticisms, and takes account of the Institute's responsibilities.

FOREWORD

Within the next year, the Congress and the Federal Executive will consider the Assessment's longer range future, including the question of its continuance. This will come at a time of heightened questioning of the Federal Government's role in education, of the institutional forms through which this role should be exercised, and of desirable levels of federal funding. This questioning is accepted in the report as a given, not as reason to compromise principle but as part of the political context to be considered. The report is addressed to the concerns and interests of key Congressional and Executive officers and their staffs.

There are obviously other groups and forces in the educational policy development process: State and local school administrators, the educational research community, the media, and the public at large. Their interest in educational measurement and their influence on educational policy development is obviously substantial. It is a subtly different question, though, whether these groups can influence significantly what happens in the crucial immediate future of the Assessment.

So far as the report speaks to the ideals and goals of educational assessment, it is addressed to all these constituencies. Its more specific recommendations reflect a deliberate effort to facilitate the discharge of authority and responsibilities by the NAEP Assessment Policy Committee, the officers of the National Institute of Education, the educational committees of the Congress, and the education, management, and budget offices of the Federal Executive. We have tried to avoid shooting arrows into the air.

In a different sense, this report is designed to meet the responsibilities imposed at least implicitly by the three foundations which initiated and have supported the project. The Carnegie Corporation, the Ford Foundation, and the Spencer Foundation have become critical and constructive forces in American education. Frederic A. Mosher, Marjorie Martus, and Marion Faldet have, as project officers, participated significantly in our planning and deliberations, without ever intruding personal or institutional views. Recognition is also made of the gracious and patient cooperation of the Assessment staff and leadership.

The staff work on the project has been almost entirely Stephen L. Koffler's. This extraordinarily competent and effective man has done, alone, what might be reasonably expected of three or four.

Our colleague Paul E. Barton has been a constant consultant and advisor. His thoughtfulness is much of our strength and conscience.

Anne Rogers has straightened out our thinking by her incomparable editing of our copy, and Arlene Huff has performed miracles of perfection with her typewriter.

Willard Wirtz
Archie E. Lapointe

INTRODUCTION

INTRODUCTION

For twelve years now, the country has been receiving National Assessment of Educational Progress reports on the performance of nine-, thirteen-, and seventeen-year-old students in Reading, Writing, Mathematics, and seven other subjects. These reports are based on a highly sophisticated sampling of young people throughout the country. The program is carefully administered. In the judgment of experts, this is one of the most dependable measures of nationwide scholastic achievement.

Yet among a hundred Americans, most of them acutely concerned about what is happening in the nation's schools, no more than one or two have even heard of the Assessment. Its reports flicker like matches in the media and then go out. What impact the Assessment has had on educational policy or practice remains relatively obscure.

What explains this anomaly? Why is there such a gap between the Assessment's repute among experts and its apparently negligible influence? Is it the consequence of poor dissemination and public relations practices that could be corrected? Is it, rather, the blunt instruction of twelve years and $64 million of experience that the Assessment is esoteric, too refined conceptually to get its message through to a public that reads on the run or to educational administrators caught in the toils of rough political process? Or, is there a prospect of constructive remodeling of the Assessment that will bring its use closer to the level of its quality?

The questions come at a time when the national temper regarding education is short. Unflagging interest in improving young people's learning opportunities has sustained and distinguished the American tradition. But today's situation is different. As perhaps never before, the public attitude has gone beyond interest in improvement to alarm that education has deteriorated.

INTRODUCTION

There is little inclination to talk about refinements which might make a good system work better. The question about the Assessment is where it ranks on a list of priorities, not for fine-tuning education but for getting it back on the track.

The Assessment faces an additional challenge. Created in the mid-1960s as an initiative of the Federal Government, it continues to rely on federal financing. The currently dominant view is that there has been too much Washington in American education. The Assessment's future depends on its consistence with a policy of putting responsibility for strengthening education squarely on States and local communities.

There is also, though, another new factor in the equation. In fairly extraordinary fashion, a national decision has been taken to "establish and raise educational standards." This is not just reaffirmation of the traditional purpose to improve education. The new insistence is that there be specific measures of educational attainment, which will permit determining whether individual students and particular school systems are passing or failing. The relationship between this new concept of "standards" and the "assessment of educational progress" is obvious.

This "assessment of the Assessment" is put directly into the context of the new but widely accepted demand for "higher" educational standards. Consideration of the specific questions customarily raised about the Assessment's design and procedures is deferred until the issue of its sufficient reason for being, in a standards-oriented system, is faced and answered. The details of its history and what it does are in Appendix A, to be consulted as they become relevant. The controlling questions are whether the essential features of the Assessment are important to the development of higher education standards in the new sense of the phrase, and if they are, how traditional Assessment practices can be reshaped to serve their purpose.

PART I

The Assessment And Educational Standards

EDUCATIONAL STANDARDS

The new significance of educational standards in this country affects the role of the National Assessment of Educational Progress (NAEP) so materially that some preliminary consideration of the concept is essential to an appraisal of the Assessment's future course. A critical change has taken place in national thinking about the relationship between the schools and the community at large. But the terms of the new relationship are unclear. So is the meaning of "educational standards."

Although "high educational standards" have always been an article of national faith, the term stood historically simply for a broad belief in quality and excellence. It meant doing the teaching job well. The education system was confidently assumed to have within itself everything necessary to assure this. Any suggestion that explicit measurements of student performance be mandated by law, and that all students be judged according to these measurements, would have been rebuked as threatening to the ideals of educational diversity, individualized instruction, and academic independence.

That confident quiet lasted until the middle of this century. In the early 1950s Arthur Bestor made two broadside attacks on the schools, and Rudolph Flesch asked, to general public applause, "Why can't Johnny read?" Then in 1957 the Russians beat us into space. Taking Sputnik as an indictment of American education, the nation reacted quickly. The National Defense Education Act (NDEA) of 1958 put substantial amounts of federal money into education and directed its use in curricular areas determined as a matter of national policy to need strengthening. The following year, James B. Conant broadened the indictment and provided a bill of particulars in his "first report to interested parents" on The American High School Today.

Responses to the NDEA and to the Conant report reflected a new consciousness among teachers that business as usual wasn't going to be enough. Curriculums were reviewed and revised, new teaching methods introduced. "Raising educational standards" was taking on new and more specific meaning, and the sense developed that there was need for fuller public understanding of what was going on in the schools.

This was the situation in the middle 1960s, when the architects of what is now the National Assessment of Educational Progress sat down at their drawing board. They knew how severe the criticism of teachers and schools had become, and that further reaction could be expected. Anticipating the demand for better information about the educational condition, they also recognized the risk that a "national assessment" might be misinterpreted. Special precautions were taken against its appearing a step, taken under Federal Government auspices, toward "standardizing" American education.

Yet even as those draftsmen gathered, additional forces were shaping that would quickly change the tide of national thinking about educational standards, even the meaning of the term.

EDUCATIONAL STANDARDS

That was the time when the high schools were getting caught up in a confluence of three ironically coincident pressures. Required to accommodate the high influx of post-World War II progeny, to retain a higher percentage of young people on through to high school graduation, and to eliminate immediately all vestiges in the classrooms of two centuries of racial bigotry, the schools were becoming crucibles of social turbulence.

The triple assignment couldn't be met, at least with the resources at hand. The Elementary and Secondary Education Act of 1965 allocated additional funds, to be concentrated on economically distressed areas of the country. New federal programs were set up: Operations Head Start and Follow Through, Learning Centers, tutoring projects, Title I (ESEA) programs. New schools were built and more teachers hired to accommodate the additional millions of students. But the citizen role was conceived of largely in terms of picking up an increased bill. Beyond that, responsibility for meeting the problems was left to government and the profession.

By the early 1970s, however, the national sense developed that educational quality was deteriorating rapidly and dangerously. It is not entirely clear how far this went, how general it was, or what measure of fault was the schools'. But the judgment was made clearly and firmly.

The reaction to this alarm took a dramatically new form. For reasons rooted in other developments, many people had lost confidence in the government and in the professions. The strongly sensed deterioration in education seemed confirmation of the failures of both of these services. So it was decided not to rely this time on either of these agencies. The decision, instead, in one State and community after another, was to move in directly on the schools, not with funds but with *"standards."*

Thirty-nine States adopted, in many instances by legislation, "minimum competency" testing programs. Standardized tests were either developed by state education agencies or obtained from commercial publishers. In many cases specific scores on the examinations were set as marking the lowest levels of competency that could be considered acceptable. The tests were given to students at specified grade levels. Often, students failing to come up to the minimum were required to take remedial courses.

Although these minimum competency testing programs represented the most extreme form of community reaction to the crisis in the schools, other patterns of response may well have broader long-range significance. They all involve some form of participation by representatives of the total community in deciding what the schools' standards are to be. These standards are applied to all students. They are also taken as measures in determining the schools' accountability to the community. This is drastic surgery on the traditional concept of relationships between the two.

EDUCATIONAL STANDARDS 4

It would be plainly wrong to imply by so summary a review that this change took place without dissent, that its rapid momentum proves its validity, or that its future course has been irrevocably set. None of these things is true. The record is that a far-reaching decision about a long-standing, deep-rooted issue has been made suddenly and firmly, but with little attention to critical details and even less to operational implications. Both the reach of the decision and the significance of what remains to be settled bear directly on the role of the National Assessment of Educational Progress.

The Assessment's genesis and subsequent history are marked by evident concern about going beyond both public and professional receptivity to the idea of reducing educational accomplishment to numbers. Whatever basis that concern may have had is now gone. The setting of standards in the current sense requires identifying and measuring student achievement levels and reporting them in mathematical terms. There is no longer any question about whether there are to be such standards. There are to be.

The dispute about public accountability is also over for all practical purposes. Teachers and school administrators continue, with considerable reason, to protest against being held solely accountable for difficulties and failures to which parents and communities are at least accomplices. But the new rule is that the political community is to be represented in establishing the standards, and that the results of the measurements are to be reported in terms that parents, taxpayers, and people generally can understand.

This confirmation of the value of the educational assessment function only opens the door to the critical questions the standards decision leaves for resolution. It is clear only in principle. There are differences as to what "standards," or "higher standards," mean. Agreement on public accountability doesn't cover critical details of how the accounts are to be set up.

Fitting the National Assessment into this picture requires coming to terms with the definitional problem, but with due regard for the fact that part of the force behind the educational standards concept is that the key words appeal to so many different people for so many different reasons. They attract a coalition constituency: the tuition poor father of a college sophomore still taking remedial reading because her high school let her off and out too easily; the highly educated mother concerned about her son's graduating from high school without a single course in either a dead or a live foreign language; employers who are looking almost frantically for young people with scientific and engineering training, and others who can't get part-time help able to meet the barest minimums of dependability. Include, perhaps most significantly, teachers who have been wanting for quite a while to get back the authority

EDUCATIONAL STANDARDS

they used to have to flunk a lazy student and make it stick. "Improving education by raising educational standards" is a compelling idea to a clear working majority of people who have very different notions about what those standards should be.

Probably most members of this coalition constituency think of the "standards" involved here as being the grades students must make on tests in order to pass. "Higher educational standards" or "raising educational standards" implies to them simply moving the passing grade up whatever number of notches is determined to be available. It may also mean tightening up on what it takes to get an A or a B-plus.

Regardless of its currency, this narrow definition does serious disservice to the "standards" concept. Its infinitely larger potential for improving education lies in developing measuring systems that are carefully tuned to whatever are determined to be the purposes of training students in particular areas. It would be reckless to raise the passing grade on an examination that measures the wrong things.

An "educational standard" is taken, for purposes of this report, as being the measuring instrument or process used to determine the level of students' educational achievement. Setting a standard includes determining how the responses to all the items or exercises that make up the instrument are to be scored, and what significance is to be given various scores. A "better" educational standard is one that measures and reports more accurately on what are rationally determined, in constructing the standard, to be the critical elements of desired student achievement. A standard can be made "higher" either by improving the educational objectives on which it is based or by raising scores or grades that are required for "passing" and for being considered superior or excellent.

This, then, is the nexus of newly made decisions and newly created problems, of definitional complexities reflecting constructive political ambiguities — all pertaining to educational standards — in which the functions and values of the National Assessment of Educational Progress must be evaluated and recommendations regarding it made.

The easiest appraisal of the National Assessment would put exclusive emphasis on three factors in this situation:

First. The Assessment has not, so far, done well in the market of public notice. After twelve years of annual announcements of Assessment findings, very few people even know about it.

Second. Most states have now set up their own minimum competency testing programs, or educational assessment systems, or both. A substantial number of commercially developed standardized tests are available. Students are already spending a disturbing amount of time being tested to find out how the educational system is working.

Third. The National Assessment is a federally financed program. The movement today is

EDUCATIONAL STANDARDS

toward reducing the Federal Government's role in education. The NAEP program has already been cut at least in half because of rising costs and previous Administrations' sharp cuts in NAEP appropriations. It is presently operating at a level below marginal utility.

These are formidable arguments for the negative. If the question were whether the National Assessment should be maintained in its present form, or even with minor modifications, the answer would be No. But that question would be too narrow and that answer irresponsible.

Despite this situation, the National Assessment provides these elements:

● It measures educational achievement among students throughout the country, using a sampling technique uniformly recognized as superior for this purpose to any other.

● These measurements are taken in different subjects, so far as available funds permit, at either four-year or five-year intervals, and the measuring is done on a basis that permits reliable identification of nationwide changes from one "assessment" to the next. There are now data available in the Assessment permitting comparison of Writing achievement results in 1970, 1974, 1979; Science results in 1970, 1973, 1977; Reading and Literature in 1971, 1975, 1980; Citizenship and Social Studies in 1970, 1972, 1976; Mathematics in 1973, 1978; Music in 1972, 1979; Art in 1975, 1979; and Career and Occupational Development assessment was made in 1974.

Most of this report is taken up with proposed improvements in the National Assessment, relying on the instruction which experience in administering it offers. But these recommendations proceed from a central conclusion that the essential elements of the Assessment are critically important to the effective implementation of the new educational standards policy. These elements are appropriately noted in three groupings, involving the Assessment's public reporting function, its contribution to the development of higher education standards, and its services to other educational agencies.

Report Card For A Nation

Nothing in the adoption of an educational standards policy diminishes the recognition that the place to improve formal education is in the classroom and that what is done there is a local responsibility. Yet there is no question about the constructive effect of concentrating national attention on a common problem by reporting the situation in the country at large.

EDUCATIONAL STANDARDS

Headlines about the nation's health have undisputed impact: "High Blood Pressure Major Nationwide Cause of Death"; "Cigarette Smoking Contributes to Shortened Life Span of Nation's Men and Women"; "National Statistics Confirm Breast Cancer No. 2 Problem of Women Between 30 and 50." The effect isn't solely or even principally on federal legislation. These headlined impersonal statistics prompt millions of individuals to do things they wouldn't otherwise have done: get checkups at the doctor's office, ease up a little, stop smoking. National statistics result in communities, corporations, labor unions, and private service organizations sponsoring blood pressure clinics, x-ray programs, and break-the-habit clinics. High schools and colleges add preventive medicine courses to the curriculums.

These statistics may tell people little they hadn't already known or sensed intuitively. Their power derives from the psychology that attaches to broadly based statistics: "If this is true of so many people, the odds are stronger that I'm one of them."

The same thing happens in other areas. Ever accelerating national youth unemployment figures are read with alarm in an economically healthy Houston and a depressed Detroit alike. Increased attention is then given in both cities to local career education, guidance and counseling, and education-work council initiatives.

It is pertinent that the events stimulating changes in the educational standards concept were national in scope: reaction to the Sputnik, the sudden and huge increase in student numbers, a more critical attitude toward dropping out, and the demand for educational equity. The report found "social dynamite" in schools throughout the country. Probably very few States would have adopted minimum competency tests if national attention hadn't been centered on the spectre of "functional illiteracy" and figures in the hundreds of thousands quoted to bring the spectre close to home.

Whatever question there may once have been about the relevance of nationwide statistics to educational policy has disappeared in the attention given during recent years to the decline in the college entrance examination scores. The sponsors of those tests warn against the error of relying on them as overall measures of the educational condition. Reporting on a self-selected and changing group, those going on to college, they are designed only to predict how well those students may be expected to do in their first year of college. But that warning gets lost in people's eagerness to find confirmation in national statistics of the persuasion that education has become weaker in their own communities.

Experience has unquestionably sobered earlier expectations regarding the National Assessment's effectiveness as an instrument for informing general public opinion. This remains for further consideration in connection with Part II. But one of the unquestionably important

EDUCATIONAL STANDARDS

elements in the Assessment program is its capacity — however incompletely it has been developed — to provide reliable and dependable information on what is happening to educational achievement throughout the country as a whole.

The Assessment developed, in the mid-1960s, from the thought that something of this kind was necessary to permit the federal Commissioner of Education to meet the statutory mandate that an annual report be made to the Congress on the condition of education in the United States. This seemed more important fifteen years ago than it does today. The current Secretary of Education, himself a former Commissioner, has recently endorsed the Assessment program strongly. His pointed advice is that it be directed toward educational policy development not at the federal but at the state level.

This could not properly be taken to suggest that federal education officials, executive and legislative alike, don't or won't have a substantial interest in nationwide educational achievement levels. Annual federal expenditures for education remain, despite curtailments, in the billions of dollars, and block grants make the kind of measurement the Assessment provides — of the outcomes of education — increasingly valuable. It remains important, both politically and administratively, that the Assessment's principal clients are not the Federal Government but, rather, whatever is meant by "the public" and, so far as government offices are concerned, those at the state and local levels.

The Assessment's value as an instrument for informing national opinion obviously depends on its quality and on its providing information not available elsewhere. The Assessment's distinguishing characteristics are the scope of its coverage (students in the elementary and secondary schools in the country as a whole) and the breadth of its objective (to measure educational achievement in the broadest responsible sense of that term). Other systems are set up to accomplish various parts of this overall task. Although the more limited information they produce can be pieced together, only the Assessment takes the broader purpose as its assignment.

The Assessment also identifies changes in student achievement levels over periods of time, characteristically four-year or five-year periods. There are some who challenge the reliability of such comparisons and others who question their value. Whether something is better or worse than it used to be seems less important than whether the present condition is good or bad, satisfactory or unsatisfactory. Yet the popular attraction of comparisons is plain, and the Assessment's titular claim to measuring "Progress" reflects a widely accepted value.

The Assessment is recognized as a model of statistical integrity and careful instrument administration. Its nationwide student sample is developed by a

EDUCATIONAL STANDARDS

combination of three-staged stratified probability sampling and matrix sampling (described in Appendix A) which is accepted as being of superior quality. The 300 to 700 items included in each Assessment are divided up into packages (usually of about forty items) and each participating student is given one package, requiring only an hour of her or his time.

The instruments are administered by specially trained personnel who are sent to the schools for this purpose. They use audio-tapes to pace the actual administration. Student responses are entered in the item booklets instead of on separate answer sheets. Where budgetary constraints have permitted, different kinds of items have been used: some multiple-choice, some open-ended written exercises, some individually administered "hands-on" or oral exercises.

The National Assessment provides a painstakingly prepared report card on the academic achievements of students at three age levels throughout the country in ten different subjects and on a basis permitting comparisons over time. This is an invaluable asset in the establishment of a system of educational standards and of public accountability.

Higher Educational Standards

The National Assessment is a good deal more than a report card. Its essential character, and in the end its reason for being or not being, lies in the quality of the value judgments which are made in the development of the Assessment instruments. Little realized by the general public, this is the heart of any rational concept of higher educational standards.

Measuring student achievement is an entirely different business from measuring other aspects of the national condition. The keepers of economic indicators — measurers of rates of inflation and unemployment and the growth of the gross national product — start from clear and precise definitions of the elements of what they are measuring. They make head-count or dollar-count surveys, compile the data, and announce the results with decimal-point precision. They get to their answers without having to make value judgments. Not so of the measurers of "educational achievement." The key term isn't defined except as they develop its meaning. Once that definition is worked out, the measuring process depends at critical points on what are in significant part value judgments.

Whether an educational standard is "better" or "higher" depends on how it consists with ultimate educational purposes. This question comes into play when a decision is made about what subjects or learning areas are to be singled out for measurement. In an even more subtle form, the character and effect of an assessment or a standard are

EDUCATIONAL STANDARDS

determined when the "objectives" within a learning area are identified for purposes of the measurement, and when items are developed for inclusion in the instrument to test students' attainment of these objectives. Another sensitive judgment has to be made in deciding how "scores" are to be worked out and what effect they are to have.

Consideration of the National Assessment's contribution to making these critical value judgments is reserved for fuller treatment in connection with Part II, for this is a matter of developing a potential as much as recognizing superiority achieved. But the Assessment's capacity for making a special contribution is manifest.

The concentration of attention by other educational measurement systems on the three basics gives the National Assessment's coverage of seven other subjects increased significance. Widespread emphasis on minimum competencies increases the importance of the Assessment's regard for all levels of achievement, from functional literacy to excellence. There are signs in the most recent National Assessment reports that the objectives on which it bases its measurements are to be drawn not from standard practice in the schools but from a consideration of how those practices can be improved.

Those in charge of the Assessment are in a position to guide their policies entirely by a determination of whatever "quality" means. They face no competition and are subject to no political pressures. Innovation and experimentation are part of the Assessment's authentic tradition. It can provide not only special competence but conscience and courage in the implementation of the new national purpose to improve educational standards.

A Service Agency

It in no way diminishes the Assessment's importance as reporter to the nation to emphasize its equal value in providing services to other agencies in the educational system. The opportunities for this are clearest in connection with the functions of state and local offices charged with responsibility for administering their own minimum competency or assessment programs. The prospects here have already been referred to in general terms and will be developed in more detail in connection with Recommendation II-I.

It remains to note, as preface to further development in Parts III and IV, the broader potential value of the Assessment to those in charge of educational policy development at the state and local level, and to teachers. The critical element in the development of this potential involves supplementing its broad assessment with additional measurings more sharply focussed on areas and subjects of special concern.

The point will be made that the energies of chief state school officers, local superintendents, and principals are consumed by pressing concerns very different from any the Assessment is likely to address: maintaining revenues, for example, and shutting down school buildings,

EDUCATIONAL STANDARDS

handling disciplinary problems, controlling drugs and alcohol and pregnancies, and on and on. It may also be suggested that these officials face such congeries of political pressures of one kind or another that Assessment reports on the state of educational progress in the nation will be quickly dismissed. The Assessment plainly is no part of an educational administrator's basic survival kit. If its product is hopelessly beyond the reach or use of even the competent and conscientious school official, the effort will be pointless. But any assumption that the state and local educational policy development process is itself non-educable is absurd.

It will be proposed in Recommendation III-B that a system of specific assessments be developed to supplement the Assessment's traditional program, and that these be designed to illuminate particular educational policy issues, such as the various issues of educational equity. So would the problem of raising minimum standards without compromising the broader pursuit of quality. Innumerable curricular questions continue to both plague and stimulate the responsible and humble administrator. If many of these questions and issues demand a different kind of analysis than the typical assessment process permits, the fact remains that such analyses can often proceed more effectively from a broadly assembled data base.

A degree of skepticism properly attends consideration of the Assessment's being directly helpful to teachers themselves. Only a scattered few teachers even see Assessment reports, and most of those don't have time to translate their implications into terms they can apply in their classrooms.

This is the eternal question facing any affirmative, quality-oriented educational initiative. The Assessment's influence on teaching can only be channeled into the entire system through which this professional competence is developed and maintained: college and university departments of education, teacher supervisory agencies in local systems, curriculum development offices, professional educational media, and associations of teachers in particular disciplines.

Later in this report, recommendation III-A is directed to improving the Assessment's utility as a basis for analysis and research. Its architects and administrators have emphasized that its purpose is to measure the results or outcomes of the educational process - as distinguished from trying to identify the factors contributing to these results. This had led, unnecessarily, to a failure to provide ready access to the Assessment data.

This data-base now contains the item responses of more than 900,000 students at three age levels in ten learning areas over a twelve-year period. Procedures are currently being developed for making these data easily accessible for secondary analysis. This will open up an important dimension of the Assessment's value.

Recommendation I: That the essential elements of the National Assessment be maintained

EDUCATIONAL STANDARDS

as vital factors in implementing an educational standards policy: by reporting nationwide student achievement levels; by showing changes over time; by developing objectives and exercises consistent with effective educational quality and process, and; by providing services to state and local assessments, testing, and standard-setting agencies.

Summary

The first recommendation, that the essential features of the National Assessment of Educational Progress be maintained, is in a sense only a preface to those that follow.

The recommendation proceeds from the recognition that if the question were whether the Assessment should be continued in its present, or even some slightly modified form, the responsible answer would be No. It has not developed an effective form of communicating the information it collects. Other educational measuring systems are being relied on increasingly. The Assessment funding has been cut back to a point that carries the program below the level of marginal utility. So the operative recommendations here are those that follow, for major changes in the Assessment program which will give it a vitality and effectiveness it currently lacks.

But the recommendation that the Assessment's essential features be maintained is made with deep conviction. This is a program currently in place which has built into it values that are essential to implementation of the educational standards concept. These values include a method of reporting the educational achievement levels of students throughout the country. Its sampling technique assures its conveying an accurate picture. Extraordinary care is taken in administering the measuring instruments. The Assessment also permits gauging changes in this condition over intervals of time. Its present data-base includes significant information covering the past twelve years and all of the principal elementary and secondary subject areas.

The Assessment's even more critical values lie in its potential for developing the skill of responsible measuring of the intangible called educational achievement. Educational standards will be only as good as the objectives which are identified and the exercises which are developed for making the measurements. If these are technicalities they are nonetheless determinative of the validity of educational standards.

The Assessment by no means reflects perfection of these measuring skills. It has carried them forward. Its position provides superior opportunities for both developing these skills and putting them to their highest uses. Its administrators have in the past pioneered and innovated and experimented. They are in position to provide critically important service functions to state and local educational testing and assessment agencies, educational administrators, and teachers.

The Assessment represents an underdeveloped and so far under-used national asset of infinite potential. To throw it away would be reckless. The critical questions are about how to improve it and increase its effectiveness. Turning, then, to those questions ....

PART II

Perfecting The Assessment

THE ASSESSMENT

It would be pleasant to think that critics of the National Assessment, trying to help chart its future course, might draw from the same wellspring its founders did. This would mean unswerving commitment to the highest ideals of educational quality, disciplined recognition of how much these ideals ask of those who measure education's accomplishments, and a strong bias toward innovation.

Francis Keppel, Commissioner of Education in the middle 1960s, knew it was ahead of those times, but only a little, to call for the measurement of student achievement through the country as a whole. The Carnegie Corporation and the Ford Foundation invested deliberately in the future. Ralph Tyler undertook the Assessment's designing as one more bold enterprise in a remarkable career which continues to be distinguished by a sense of responsible adventure.

This tradition invites criticism and depends for its renewal on change. What was innovation in educational assessment ten years ago is now custom and habit. If this gives the Assessment's architects satisfaction, it must also make them restless. The only approach to the Assessment's future consistent with its authentic character is to ask what the educational measuring process most needs today and is least likely to get anyplace else.

The Assessment's current stewards ask no quarter as far as criticism is concerned. To the contrary. Many of the proposals which follow here come from NAEP officers and staff. Because the Assessment budget has been cut back sharply over the past five years, contemplated innovations have been sacrificed to meet survival's demands. Many of those closest to the Assessment agree that continuing it on the current basis would not make sense.

Small suggestions will be wasted except as they are part of a proposal sufficiently broad to justify the Assessment's revitalization. The conclusion that its essential elements are critical to the development and implementation of a new educational standards policy leaves the question of whether the Assessment program can be expected to develop these essential elements in a manner and to a degree that will make its promise unique. Unless such a prospect can be marked out, these elements should be bequeathed to other educational assessment agencies.

The specific recommendations which follow here are based on a dual conviction: that an educational standards policy depends for its effectiveness on the perfection of the "value judgment" elements in measuring educational achievement; and that this can be best attempted by an independent agency working from a nationwide base.

Seldom recognized by "the public" which is demanding "standards" and "accountability," these value judgment elements can be given short shrift by the measurers. Limited funding may be a constraining factor, especially when the need to experiment is taken into account. Pressure to produce both

THE ASSESSMENT

quick and encouraging results is likely to compromise efforts to build in quality at points where it isn't going to show.

None of this is meant to reflect on other educational assessment or standard-setting programs. Understood properly, it simply recognizes that some of these processes can be developed more effectively in a laboratory than on the firing line or in the marketplace. This principle is completely accepted for the physical sciences, and the benefits accruing to those in the market and on the battlements are recognized. The principle is no less sound with regard to the social sciences, including the science of measuring educational achievement. The Assessment program — or whatever is established in its place — will be properly perceived as in part a laboratory, a place for scientific invention, where inspiration is loyal to quality and free from diverting pressures.

The most convenient approach pattern for considering how to improve the Assessment is to take up the various functions in roughly the order of their effect on the construction and administration of an assessment program.

Assessment Coverage And Frequency

A preliminary but basic value judgment affecting education's character is made when certain aspects of it are singled out for measurement. We tend to do best those things we measure most closely. We become to a considerable degree what we measure.

The original National Assessment plan was to take periodic measurements in ten academic subject areas. This decision was made after deliberately weighing the value of measuring subjects traditionally emphasized in the schools against the importance of finding out what is happening in areas that receive less attention but may be equally important to learning's highest principles and purposes.

The first surveys, started in 1969-70, were of Writing, Science, and Citizenship. The results of these first assessments were reported publicly in 1970. An average of two assessments each year was contemplated.

Over the past twelve years, some twenty assessments have been completed and reports issued: in Science, Writing, Reading, Literature, Citizenship, Social Studies, Mathematics, Music, Art, and Career and Occupational Development. In 1977, an adult group was assessed in Health, Energy, Reading and Science. Occasional special assessment probes have been made in basic mathematics, basic life skills, and consumer skills.

A combination of reduced funding (from $7 million in 1973 to slightly less than $4 million in

THE ASSESSMENT

1981) and increasing costs has resulted in a serious narrowing of Assessment coverage. The present funding level permits making only one assessment each year, and a 1978 amendment to the authorizing legislation requires measurements at least once every five years in Reading, Writing, and Mathematics.

It would be preferable to measure these basic subjects at four-year rather than five-year intervals. This would permit closer watching of the effects on educational results of the emphasis being placed on standards. It would also permit some degree of "longitudinal" analysis of the results of surveys given the same student generations at the nine-year, thirteen-year and seventeen-year levels. (This is considered further in connection with Recommendation III-B.) Yet if a choice must be made between limiting the Assessment's subject coverage and extending the time between measurements, the latter is clearly preferable.

From one perspective, the effect of statutory instruction and these funding limitations is to leave the Assessment program in line with the current nationwide emphasis on what is happening in the three basic subjects. But increasing concern is expressed about the possible side effects of concentrating attention solely on the educational core, and there is widening insistence that the importance of the arts and sciences and broader humanities not be downgraded. Part of the value of the National Assessment has been its broad subject coverage.

Reducing the NAEP program to one assessment each year, confined largely to Reading, Writing, and Mathematics, poses squarely a critical question of marginal utility. Such a schedule will virtually assure the Assessment's not playing a major role in informing the general public about the educational achievement picture. Other reports — of state or local measurings and of the results on college entrance examinations — come out annually and are widely covered in the media. The occasional appearance of additional information from a little known source will seem to reporters the stuff for footnotes and inside pages.

This leaves the question of whether the Assessment would still have significant value as a program for developing new and higher fidelity measuring and assessing and standard-setting techniques and materials which state and local offices could use in developing their own tests, assessments, and standards. The Assessment's own occasional measurings would provide a basis for checking the results of other measurement systems.

These are not negligible values. Assuming the development of unique technical competencies, they are critically important factors. But if no significant public reporting element is to be included in the Assessment's functions, it will be questionable whether a separate agency should be maintained for these purposes alone. Alternative arrangements would probably be better made for developing these capacities in connection with other assessment programs. Yet the losses resulting from taking

THE ASSESSMENT

that alternative course would be substantial.

No precise computation of cost effectiveness, public recognition factors, and marginal utility is possible here. Assessment frequency is important but of secondary significance. Covering subject areas not covered by other assessment systems is critical. With overhead costs remaining virtually constant, two assessments per year cost about half again as much as one does. The continued vitality and economy of this program depends on its producing at least two assessments each year, including to some extent the "specific assessments" proposed in Recommendation III-B.

Recommendation II-A: That the National Assessment program be restored to a basis permitting at least two assessments each year.

Sampling And Instrument Administration

Extraordinary measures are taken to assure both the representativeness of the nationwide sample selected for making the Assessments and the efficiency and uniformity of the practices by which the instruments are administered. The Assessment's statistical integrity is one of its distinguishing characteristics, to be fully respected in any consideration of possible modification — but not without regard to several questions these high fidelity practices raise.

The cost factors which are involved are substantial. Some close observers of Assessment operations insist that these practices have been carried past the point of diminishing returns, and that the Assessment's refinement to meet statisticians' precepts has limited its extension to cover important policy needs.

The NAEP design involves two types of sampling: three-staged stratified probability sampling and matrix sampling. The details of the design are spelled out in Appendix A. They give priority to three considerations: the necessity of including enough representative students to provide a reliable nationwide sample; the desirability of including a sufficiently large number of items to cover a broad spectrum of curriculum elements in the subject being assessed; and the administrative decision to limit to one hour the amount of time required of a student participating in the assessment. These purposes are accomplished by dividing the items making up each assessment (300 to 700) into a number of different packages (typically eight to sixteen), with each student being given only one of these packages (usually containing forty to forty-five items). From 2,000 to 2,500 students are normally given each package.

THE ASSESSMENT

Sending out specially trained NAEP personnel to administer the instrument to students is the most costly instrument administration practice. Audio-taped instructions are also used to pace the student response process. Certain other precautions are taken to assure the highest possible degree of uniformity in various aspects of the administration.

Review of the NAEP sampling design confirms its importance to satisfying the NAEP objective. It is expensive. It is also essential to the integrity of the National Assessment measurements.

Accepting the established NAEP policies regarding the development of an adequate student sample, an important and related operational question remains of how many achievement items must be included in the assessment to permit a sufficiently comprehensive and detailed coverage of the subject being surveyed. This is determined in practice by balancing the recommendations of the Assessment Advisory Committee against budgetary limitations. The question of how many "student background variable" items are to be included also comes into this equation. The traditional one-hour limitation on the amount of student time which can be used has an important effect. The ultimate determination takes the form of a decision as to how many packages of items must be given to how many students.

A review of these practices confirms the responsibility which is being exercised in keeping the number of students included in the sample and the number of achievement items to minimums consistent with the demands of adequate subject-matter coverage and the integrity of the assessment results. In recent assessment administrations, fewer packages than before have been given to fewer students, and the effect of this on the integrity of the final assessment results are being carefully watched. The only remaining question about item inclusion, involving the balance between achievement items and student-background variable items, is discussed further in Part III of this report.

One relatively minor recommendation is made so far as the background variable items are concerned. Students' achievements at particular age or grade levels are bound to be affected by the degree of their exposure to the subject being assessed. Current NAEP practice is in fact to ask students participating in the assessment how many courses they have had in the subject. Factoring this information (along with the student's age) into the publicly reported results would increase their value.

The limitation on the use of student time to one hour warrants further consideration. Increasing the time period to two hours would permit major savings which could then be put to increasing the Assessment's utility in a variety of important ways. These are only illustrated by suggesting the possibility of using the same student sample to conduct assessments in two different subjects, spending the first hour on one, the second on the other.

THE ASSESSMENT

Flexibility in working with more student background or educational context items would obviously be increased. A fuller basis could be established for secondary analysis of the assessment results.

Three restraining considerations are properly given account. The effect of student fatigue on examination results is debated endlessly. NAEP's architects were also seriously concerned about whether local school authorities would cooperate on a more than one hour basis. The American Psychological Association's ethical code properly directs critical attention to any use of student time for testing or assessment which does not serve an educational function for the students themselves.

It can only be a value judgment that the operational advantages of changing to a two-hour system and the values of the National Assessment outweigh arguments to the contrary. Almost all other standardized examinations and assessments are now administered on a two-hour-or-more basis. Various states and local systems are now relying on NAEP services and the results of the National Assessments to an extent which would appear to assure their cooperation. If the National Assessment contributes as significantly to the setting of quality educational standards as its potential promises, its value to students will be clearly identifiable. It is recommended that this important operational change be made.

Another close question carrying a substantial price involves the practice of using only special NAEP-trained personnel to administer the instruments to students. This means maintaining twelve District Supervisors and hiring about 150 Exercise Administrators for two-week periods in connection with each assessment. Most state and local assessment and minimum competency programs rely for this function on teachers or other personnel from within the local school system. So does the Assessment of Performance Unit (APU) system, the British equivalent of the NAEP. This administration factor is deemed so important by Assessment officers and staff, and by most observers, that recommending a change in it would risk irresponsibility. The cost factor is large. It is not clear that full exploration has been made of alternative possibilities.

Recommendation II-B: That current Assessment sampling practices be maintained; that the one-hour limitation on the use of student time be increased to two hours; that continuing attention be given to possible economies in instrument administration practices; and that the Assessment reports include information regarding the courses students have had in the subject area assessed.

THE ASSESSMENT

State-By-State Reporting

National Assessment sampling policies also bear directly on the suggestion frequently made that Assessment results be reported on a state-by-state basis, at least where this is requested by the officers of a State. The short answer is that the NAEP sample is not large enough to permit this. It does allow the Assessment results to be reported for each geographical quarter of the country. Analyses could also be made from the present sample of possibly significant groupings of States, for example, on the basis of high and low per capita income. But expanding the Assessment to permit reliable reporting on a state-by-state basis for all States would require a sample several times the present size, and would increase the cost of each assessment far beyond practical limits.

This short answer is neither complete nor sufficient. Giving the Assessment this comparison capacity would greatly increase its utility. It would help meet the statutory mandates in several States for a public reporting of how the statewide educational assessment results measure up to national averages. Whether wisely or not, these comparisons have become a major factor in implementing the decision to give new significance to educational standards.

It was part of the original intent of the Assessment architects to establish it on a basis which would permit state-by-state comparisons. This intention appears to have been abandoned for a combination of two reasons. It became apparent that the use of the sampling design the technical advisers considered necessary would be impracticably expensive if a state-by-state breakdown was to be provided. It is also suggested in the record that there were political concerns about such comparisons contributing to the misconception of the Assessment as a step toward federal control of education.

The cost factor continues to be real, and although the political consideration has less basis today than it did fifteen years ago, it cannot be disregarded. Any effort to enlarge the Assessment program to the point that it would publish state-by-state results for all States would probably be effectively opposed.

A clear alternative course is available. Arrangements can be made for the inclusion in a state assessment instrument of enough National Assessment items to make possible a comparison of state with national results. State officers can, if they desire, use all or part of a NAEP instrument for their own assessment purposes. As will be noted later, a good deal of this general kind of use is actually now being made of the NAEP resources.

Part of the importance of this possibility lies in the prospect it presents for a supplementary source of National Assessment financial support. Other possibilities of reciprocal arrangements might involve building into the National Assessment sample some of the students assessed as part of a

THE ASSESSMENT

state program. These prospects go beyond cooperation with the States to include comparable opportunities involving local school systems.

Recommendation II-C: That the Assessment not be expanded to permit reporting of results on a state-by-state basis; but that arrangements be made to facilitate use of the Assessment by state or local school agencies for comparisons with nationwide student achievement levels.

Age Or Grade Assessment?

The value of the Assessment to state and local education offices would be substantially enlarged if its surveys were made on a grade-level instead of the present age-level basis. But there are competing considerations involved here, and additional aspects of this coverage question.

Original NAEP practice was to measure four age-groups: nine, thirteen, seventeen, and twenty-six to thirty-five. The seventeen-year-old group included both in-school and out-of-school youth. Measurement of the young adult group and inclusion in the seventeen-year-old group of those who have left school have now been virtually abandoned because of the costs involved.

These cutbacks represent serious losses to the Assessment program. Its coverage of out-of-school seventeen-year-olds and of young adults gave it unique value. Although the costs are substantial, these cutbacks are false economies.

The larger issue is whether NAEP measurements should continue to be taken at age levels, or whether it would be preferable to shift to measuring students at grade levels. School officers urge this change. Since most state and local assessments are made at grade levels, they argue that putting the NAEP on the same basis would make its data more useful and increase the possibilities of at least partial integration of the national system and state or local systems.

The original decision to administer the Assessment on an age-group basis reflected concern that differences in entrance and promotion policies might affect assessment results. It also had to do with the effort in the 1960s to get away from traditional thinking that a child's educational diet had to be determined by her or his assignment to a particular grade. The additional argument is made that to shift now from an age-group to a grade-level basis would disrupt the measurement of changes over time.

The question is close and a reasonably broad survey reveals divided professional opinion. While school administrators and teachers, in the main, urge the change to a grade-level basis, educational researchers would continue the age-group practice in order not to disrupt the current indices.

THE ASSESSMENT

The close balance of values has prompted investigation of the difficulties and costs of making measurements on both bases. The best available advice is that this would require a twenty-five percent increase in the size of the required sample, meaning an estimated $225,000 increase in the total cost of each subject assessment. This is too much.

The practical course is to change to a grade-level assessment. Data from previous NAEP administrations identify the grade of participating students, making it practicable to adjust the earlier data for comparison purposes.

One other point is worth noting. Present NAEP practice is to identify nine-year-olds and thirteen-year-olds in calendar-year terms, but seventeen-year-olds on an October 1st through September 30th basis. This means that most seventeen-year olds for NAEP's purposes are in the eleventh rather than the twelfth grade, which accentuates the difficulty of relating NAEP data to those developed in most other assessment systems. Changing to a grade-level basis would eliminate this problem, but if this change is not made, the NAEP definition of the seventeen-year-old group should be revised to a calendar-year basis.

Recommendation II-D: That the Assessment be administered in the future on a grade-level basis at the fourth, eighth, and twelfth grades; and that it include, as it did originally, out-of-school seventeen-year-olds and a young adult group.

Objectives And Exercises

A statement in the NAEP Design and Development Document covering the 1979-80 Reading/Literature Assessment is succinctly descriptive:

The first step in any assessment cycle is objectives development. The objectives identify the important knowledge, skills and attitudes within an assessment area which are generally being taught or should be taught in schools. These objectives then become the framework for developing assessment exercises which measure the objectives.

Although there is little public awareness of these steps in the process of setting educational standards, they affect that process vitally and give any standard its determinative character. They require value judgments. Responsible recommendations regarding the National Assessment must turn on how it can be expected to handle the objectives-setting and item-selection steps of the measuring process.

The practice now is to set objectives through a consensus approach involving a large number of subject-matter experts, teachers, curriculum developers, testing experts, lay persons, and minority group representatives. Special steps are taken to eliminate racial, ethnic, gender, or

THE ASSESSMENT

socioeconomic bias. NAEP staff members participate throughout.

The 1979-80 Reading/Literature Assessment is illustrative. More than 130 education professionals and lay citizens were involved during an eighteen-month period in setting the objectives. Four objectives were eventually identified: "values reading and literature"; "comprehends written works"; "responds to written works in interpretive and evaluative ways"; and "applies study skills in reading." These were amplified to identify sixteen subobjectives. Each of these subobjectives was further amplified in one to three specific illustrative questions (e.g., "Can students infer the main idea or purpose of a text?"). This process is fully reported and the complete list of objectives and subobjectives is published.

Appraisal of the effectiveness of this process depends on which of two concepts of educational assessment and standards is adopted. The two are not clearly distinguishable and neither of them can rely on perfect logic or pretend complete rationality. Although catch phrases are misleading, short-hand is required for discussion's purposes: call one the comparison concept and the other the quality concept.

A traditional purpose of much educational measuring has been to make comparisons: of one student's performance with others', of the results obtained in one school system or State or country with those recorded in others, or of achievement levels over periods of time. These uses obviously require that the objectives identified for constructing the measurement have some reasonable relationship to educational goals. But this isn't pressed very hard. Emphasis is quickly shifted to assuring the internal coherence of the instruments. What is important when the comparison is being made over time is that the objectives and items be kept sufficiently constant that the comparisons are reliable.

This approach to constructing measurements is probably acceptable in implementing educational standards and accountability principles which also rely exclusively or primarily on the idea of comparison: not on what is good but on what is better or at least up to par. The operative principle of this concept is that competition spurs effort and that this will in itself mean improvement. If advisable, the score levels considered satisfactory can be raised.

The National Assessment objectives have been generally considered satisfactory for purposes of such comparisons. No validation of the objectives in terms of their relevance to broader educational goals has been asked and little attempted. Instead, reliance is placed on the process by which the objectives are set. The consensus approach is accepted as the best available, and special efforts have been made by NAEP staff to improve its mechanics. The consensus medium has become the message.

THE ASSESSMENT

But to settle for the comparison function is to disregard the potential of the decision in this country to develop higher educational standards. This decision is to improve the quality of education by strengthening it and elevating its goals. The idea is not just to get results with spurs or a whip, but to strengthen the horse.

Accepting this quality concept of standards, and recognizing the determinative influence the setting of objectives has on the character of these standards, leads to some difficult operational issues. How can objectives for a National Assessment be set so that they will both permit valid comparisons of achievement levels over periods of time and build in new qualitative elements? If setting objectives relies on consensus, how is a consensus for change to be developed? Most centrally, how can the measuring of achievement be carried beyond comparisons without intruding an element of value judgment antithetical to the essentially pluralistic nature of American education?

There are neither easy nor complete answers to these questions, which is why some of those who think most clearly about tests and standardized examinations and assessments and educational standards are most skeptical about their ultimate value. Yet if standards are to be given new significance and effect, this is all the more reason for trying to make the objectives on which the standards are based as fully consonant as possible with education's highest purposes. This new emphasis will mean that teaching will be increasingly oriented toward these objectives, which is good or bad depending on their quality. If these standards are to determine accountability, it is critical that their measurement reflect ultimate educational purposes rather than what might be dangerous expediencies.

The key demand on the National Assessment is that full use be made of the opportunities it offers for fitting educational assessment into the quality rather than simply the comparison concept of educational standards. It has faced this challenge from its beginning, though not as squarely as might have been hoped. A key question in the early debates was whether the basis for setting Assessment objectives should be "what the schools are teaching" or, rather, "what they ought to be teaching." The delphic decision to look at "what the schools are trying to teach" was cautious and conservative.

The adoption of the consensus process for identifying objectives was not. This would not be the way to select objectives based only on what the schools were in fact doing. Including citizen members in the consensus group, and increasing the role of "content experts," could only be explained as signs of interest in what education *ought* to be.

Yet the fairest appraisal is that throughout the first ten years of the Assessment's operation, the potential of the objective-setting process for improving the quality of education was sublimated to the emphasis on making these

THE ASSESSMENT

objectives valid for purposes of comparing achievement levels over time. A constructive tension was maintained between the two factors, perhaps more than in most measure systems. The several Writing and Citizenship and Career and Occupational Development assessments reflected a considerable degree of invention and innovation. But the others were predominantly traditional, oriented to comparison.

The 1979-80 Reading/Literature Assessment, reported this year, appears to reflect a critical change in NAEP emphasis. It embodies elements of objectives-setting that are essential to a quality concept of educational standards.

Two phrases in the design and development document passage quoted above are critical. Objectives are to "identify *the important* knowledge, skills and attitudes." This is to include those "which are generally being taught *or should be taught* in the schools." The emphasis is added, but is consistent with the original context. This statement contrasts with the 1970 NAEP description of the objectives set for the first Reading assessment. These were described as involving no "distinctly 'new' objectives," but as "restatements and summarizations of objectives which (have) appeared over the last quarter century." This was characterized as "a desired and expected outcome in that one criterion for National Assessment objectives (is) that they epitomize the prevailing teaching directions of educators." (Reading Objectives, 1970, p. 8.)

Reviewing the 1970-71 and 1974-75 Reading objectives with those identified for the 1979-80 assessments reveals an obviously sufficient parallelism for comparison purposes. Yet the differences are at least as important and marked as the similarities.

The Assessment's critics have complained properly in the past that in preserving its internal coherence and protecting its validity it ignored recognized school outcomes: "whether or not students can distinguish between analytic and political tasks"; "how to provide illustrations other people can understand." Subobjectives in the 1979-80 statement of objectives go directly to almost precisely these points.

Critics have also singled out the failure to establish patterns of relationship among NAEP objectives. If Piaget's "stages of intellectual development" and Bloom's "taxonomy of educational objectives" go beyond what is the measurers' business, they nevertheless suggest the need for some classification system for identifying the objectives on which an assessment is based and putting them in some order of skill demands. The New York State Education Department's Degrees of Reading Power standards represent a significant step toward meeting this need.

So it becomes important that the 1979-80 Reading/Literature Assessment statement of objectives says that although "this taxonomy of objectives is not meant to represent a hierarchy of subskills nor the sequence by

THE ASSESSMENT

which comprehension is achieved or is learned," it does constitute "a useful way of organizing the questions in this assessment so that they address important aspects of comprehension of a wide variety of text types." There is the further comment that "all of the objectives interact with one another; they only appear sequential and autonomous when they are described linearly .... Each objective and subobjective represents a continuum of difficulty." If these statements of purpose appear broader than the accomplishment, they nonetheless evidence new forces at work within the consensus process.

The two reports which have been issued on this most recent assessment confirm a marked change in NAEP practice and policy. The April, 1981 Three Assessment report identifies three basic Reading objectives (literal comprehension, inferential comprehension, reference skills) as representing "a crude but useful way of organizing assessment items in terms of the measurement focus of each." By October, the signs of renewed frontiersmanship were plainer. Entitled Reading, Thinking And Writing, this second report draws on the original objectives statement and the assessment data to suggest a four-step model (initial comprehension, preliminary interpretation, reexamination of the text, extended and documented interpretation), "through which comprehension would evolve." It then concludes from the assessment results that "American schools have been reasonably successful in teaching the majority of students to complete the first two steps of the model, but have failed to teach more than five to ten percent to move beyond their initial reading of a text."

Combining what were previously separate Reading and Literature assessments, and analyzing the results in terms of Reading, Thinking And Writing, are significant steps. Preserving artificial boundaries between various subjects, simply because of custom and the way teachers are trained, results in serious losses. These accumulate as history or science teachers disregard writing errors, and students' arithmetic papers are returned with teacher comments that abuse the language. Education in two of the three "basics" — Thinking makes it three out of four — is bound to be improved if advantage is taken of opportunities to teach them at the same time.

Although the Reading, Thinking and Writing report and analysis appeared long after the objectives had been set and the instruments developed for the 1979-80 assessment, it is fair to assume that these interrelationships will be recognized as objectives are set for the next round of assessments. The list of teachers and content experts will properly include some from disciplines other than the particular one being assessed. More attention will be given to identifying objectives that cross traditional curricular boundary lines.

Increasing emphasis is being placed on the process, as well as the content, elements in learning. If the "new math" initiative twenty years or so ago was in some respects abortive, it nevertheless increased recognition of problem solving as a basic skill. The recent experiment by the British Assessment of Performance Unit with alternative content-oriented and process-oriented assessments of mathematics illustrates how measurers can at least direct attention to the importance of skill development.

Implementing a quality concept of educational standards demands that the Assessment emphasize the higher reaches of student achievement as much as "minimal competencies." Its objectives are established in terms of across-the-board mastery. Instrument items are selected deliberately, to include easy, medium difficulty, and hard exercises. The "four-step model" developed in the Reading, Thinking And Writing report emphasizes the higher comprehension skills (which only five to ten percent of American students appear to be mastering) as well as the lower skills (more widely achieved by students). With concern growing lest minimum competencies be achieved at the cost of reduced educational quality, there is a reason to relate subobjectives and skill elements to levels of desired proficiency.

What is emerging is a concept of a broad and integrated spectrum of objectives identifiable with the improvement of educational quality. The purpose of making comparisons over time is met primarily by including certain items or exercises as constants in the successive assessments. But beyond this, the objectives are developed in a pattern determined by deliberate consideration of what it is that students ought to be able to do at particular points in their learning about a subject as they move toward mastery.

Where there are alternative approaches to this learning — by different emphases, for example, on content and on process — subobjectives can be identified and instrument items selected to permit measurement of the comparative effectiveness of the different approaches. Where there are clear or responsibly postulated interrelationships between different learning elements, either within a particular learning area or crossing traditional boundaries, objectives will be formulated and items developed to take this into account.

The spectrum of objectives can be developed with account being taken of the desirability of identifying whatever order of succeedingly higher degrees of skill is possible. No "hierarchies" — in the sense of one level of proficiency reflecting mastery of all those "beneath" it — appear practical. But the New York Degrees of Reading Power standards illustrate the possibilities of carrying the setting of objectives to a point permitting responsible measuring of how far a student has gotten along a line or even up a ladder of proficiency. The "four-step" model identified long after the 1979-80 Reading/Literature objectives were established suggests that a comparable approach will be taken in setting the objectives and subobjectives in the next National Assessment in this area.

It would be entirely wrong to suggest that the National Assess-

THE ASSESSMENT

ment processes have now been carried as far as a quality concept of educational standards demands. They clearly have not been. But those demands are now recognized and accepted. The Assessment's superior opportunity is to push ahead toward fuller implementation of this quality concept. The critical question, too rarely asked by the users of educational standards, is how good they are — not just for making comparisons but in terms of education's purposes. It will be in its attention to this question, at least equally with its reporting changes over time, that the Assessment's unique value will be established.

These suggestions assume continued reliance on the consensus approach to the setting of assessment objectives. This is increasingly important as emphasis is moved beyond "what is being taught" to "what should be taught." To invoke the value and importance of educational quality requires recognizing the problem of determining what "quality" is to mean. Whose idea of what should be taught is to prevail? Yet this underlying issue is new neither to the process of setting objectives for educational assessment nor to the process of democracy. The operative questions of consensus mechanics are critical. Coming down to the degree of administrative and participant responsibility that is exercised, they defy formulation.

Full account has been given to the importance of preserving the Assessment's validity as a measurement of nationwide changes in educational achievement over time. The 1979-80 Reading/Literature Assessment experience indicates that there is no serious problem here. The National Assessment can both keep the educational record and contribute to improving it. Both functions depend on the integrity and the vision with which its objectives are set.

Recommendation II-E: That Assessment practices and policies relating to the setting of objectives be directed toward including all elements important in improving education in particular learning areas, and that increased efforts be made to develop patterns of objectives that will facilitate the establishing of higher educational standards.

Developing and selecting exercises or items for inclusion in the assessment instruments is closely related to identifying objectives. The National Assessment practices are typical of those relied on in most standardized testing and assessment systems. Various kinds of development centers, conferences, and consultant groups are set up. Teachers in the subject being assessed, college educators, and curriculum specialists submit items which are then reviewed by committees of subject-matter specialists, lay persons, and minority group representatives. Fairly extensive field testing is part of the process. NAEP staff members work

THE ASSESSMENT

with the Assessment Advisory Committee in making the final selection of items for inclusion in the assessment instrument.

Although these practices probably produce items and exercises adequate for comparing students with each other or achievement levels over time, they are not tuned to, nor adequate to meet, the demands of a quality concept of educational standards. The items being relied on are ordinary. Little evidence of their validity in measuring students' accomplishment of the identified objectives is available and no sufficient capacity for determining this validity has been developed.

If the objectives-setting function is to be advanced to the level of sophistication which a quality concept of educational standards requires, commensurate attention will have to be given the item development and selection process. Efforts to develop an integrated spectrum of objectives will be frustrated except as more sensitive instrument items are also devised. The traditional NAEP practice of using a rough mix of easy, medium difficulty, and hard items will have to be substantially refined to justify fully the "four-step model" of graduated reading comprehension skills which the six NAEP staff members and consultants developed after the 1979-80 assessment had been completed. The new aggregating of NAEP item results for reporting purposes raises serious questions about whether the items were originally developed and put together with the care and the understanding of their interrelationship necessary to fully legitimize such aggregation.

The earlier NAEP tradition was to make the Assessment an instrument for innovation and experimentation in exercise development. Various individually administered "hands-on" exercises were used to inquire into students' ability to use what they know. The 1969-70 Citizenship Assessment included a group task exercise designed to determine, by observing students' group interaction, their ability to "apply democratic procedures on a practical level." The first Writing assessment (also 1969-70) introduced new techniques which have been widely emulated. The assessments were deliberately devised to include a mix of open-ended written exercises and multiple-choice items which would illuminate possible short-comings of either of these forms.

This capacity for innovation and experimentation has been lost, largely as a consequence of budgetary constraints. The new uses to which educational standards are being put and the new importance being attached to identifying more refined assessment objectives make the restoration of the inventive capacity imperative. Part of the National Assessment's responsibility is to probe into what those in the educational measurement profession know are recesses of ignorance about how

THE ASSESSMENT

to ask the right questions. It doesn't matter so much when the primary purpose of the measurement is comparison, to measure today's achievement levels against yesterday's or to plot answers — and students — on a bell-shaped curve. It becomes critical if the purpose of educational standards is to raise the level of learning.

NAEP administrators recognize so clearly the need to improve their staff capacity to perform these item development functions that pressing the point belabors what is obvious. They are currently seeking private foundation support for the development of a National Assessment research component which would strengthen its exercise development process by identifying "efficient and effective procedures and approaches for developing objectives and items," determining "the relationship between the present assessment parameters and the resulting measurement constraints," alleviating "the gap between current practices and state-of-the-art measurement techniques," investigating "the relationship between current exercise development procedures and appropriate secondary analysis of National Assessment data," and establishing "an in-service training component." This request reflects accurately a critical need.

Recommendation II-F: That the National Assessment procedures and capacity for developing instrument items be revised to meet the need for items that will measure with maximum accuracy, and in a variety of ways, students' proficiencies as they relate to identified educational objectives.

Reporting And Communication

In 1968, two years before the first National Assessment report appeared, its designers commented publicly on an issue which was giving them concern:

Some people feel that the National Assessment should provide information only, leaving all interpretation to the users of the results. Others feel that interpretation is necessary, if only to point out various hypotheses that are tenable, unless one wishes to run the risk of gross misinterpretation. This issue has not been settled yet. (What Is National Assessment? Education Commission of the States, 1968)

Thirteen years and twenty-one assessments later, this issue — information versus interpretation — still is not settled. Protracted ambivalence has unquestionably detracted from the Assessment's effectiveness. "Someone," that 1968 statement continued, "hopefully many different 'someones,' must pore over and sort out the results in a fashion that is most meaningful." The "someones" haven't materialized in the hoped for numbers, and hindsight

THE ASSESSMENT

clearly reveals the original expectation as a flaw in the Assessment's planning.

The early Assessments were reported on an item-by-item basis. Students at various age levels were asked, for example, to read passages from the Constitution, from a newspaper, from a driver's license application form. The percentage who could do this satisfactorily was reported. This was about it. The theory was that if the general public was informed about student achievement in terms of familiar tasks, "someone" would take it from there.

Other considerations affected that original decision, making it more plausible at the time than hindsight indicates. Over the years and in light of chastening experience, the Assessment practice has changed substantially. If the nature of the subject matter requires, as with Writing for instance, strong emphasis is still placed on what students do with respect to particular items. But most National Assessment reports now include virtually complete aggregations of item results.

The report of the 1979-80 Reading/Literature Assessment includes a single figure — 67.9 percent — as the average of the percentages of correct responses by nine-year-olds on each of the 57 exercises in the comparison-over-time section of the Reading assessment. This is set out with comparison figures for the 1969-70 assessment (64.0 percent) and the 1974-75 assessment (65.2 percent) — permitting the conclusion that by this measure reading competencies have increased by 3.9 percentage points over the decade.

Similar aggregations of item results are reported for each of three reading objectives: literal comprehension, inferential comprehension, and reference skills. The same pattern of reporting is followed with respect to thirteen-year-olds (showing a 1979-80 overall reading achievement level substantially unchanged during the decade) and to seventeen-year-olds (also virtually unchanged, but with a significant drop in inferential comprehension).

This significant reversal of the original item-by-item reporting practice has been generally applauded. Media coverage of the Reading/Literature Assessment dramatically exceeded that accorded previous NAEP reports. This is partly because these reports now permit comparisons of change over time — the Assessment's original purpose. The aggregation of item results also provides the broad picture that the media and public alike seem to consider more interesting and informative.

But is it? What do the figures covering 73 pages of the report on Three National Assessments Reading Changes In Performance, 1970-80 mean? What does it *mean* that if American seventeen-year-olds are asked 71 questions about reading, even assuming these are the best possible questions about reading, 68.2 percent of the answers will be correct? Is 68.2 percent a mark of gratifying success for

THE ASSESSMENT

American education or of alarming failure or of shabby mediocrity?

If nine-year-old boys and girls now answer correctly 67.9 percent of the reading questions, and the comparable number on the same questions nine years ago was 64.0 percent, this is clearly better. But is the 67.9 percent good or bad or something in between?

How much meaning is there for how many people in the reporting of percentages, a few decimal points higher or lower than they were nine years ago, for three different age groups' achievements in "literal comprehension," "inferential comprehension," and "reference skills"? Washington, D.C. parents rebelled recently when children brought home mid-term report cards in the form of four-page computer printouts. Those indicated how each child was doing on a long list of items included in an "Objectives Check-list" for each subject: whether in second-grade math, for example, the child could "write a two-digit number as a sum in which one addend is the next lower unit of ten," and whether, in reading, she or he could "identify initial consonant substitution" and "apply CVC (consonant vowel consonant) principle." School offices were deluged with demands to know "whether this means an A, B, C, D, or F."

Putting the idea of higher educational standards to work, in the full sense of standards based on rational objectives and subobjectives, is a complicated business. A public properly demanding "accountability" isn't very clear about what it wants. And as the National Assessment shifts from item reporting to broad aggregations of objectives percentages it faces a new set of communication problems.

One answer would be easy, immediately attractive, and fatal. The NAEP Assessment Policy Committee could decide that anything below 60 percent is failure, 60-70 percent a D, 70-80 percent a C, 80-90 percent a B, and anything above 90 percent an A. This would be a national, and in the second round of editorials a "federal" standard.

Recognizing the importance and the sensitivities of this communications problem, NAEP administrators took two innovative steps in interpreting the latest Reading/Literature assessment. Following up on an experiment made in the 1979-80 Writing Assessment, four pages of the Reading report are A Perspective On The Results developed by thirteen reading experts who meet together to discuss the findings. The succeeding report on Reading, Thinking And Writing a more intensive interpretive analysis by the six NAEP staff members and consultants who, on their own initiative, undertook to identify the significant meanings of the Three Assessments report and to suggest their implications for educational improvement.

The panel of reading experts point out changes and weak spots in reading achievement during the decade. The six staff members and consultants carry their analysis to the "major conclusions ... that American

THE ASSESSMENT

schools have been reasonably successful in teaching the majority of students to comprehend initially what they read and to make a preliminary interpetation but have failed to teach more than five to ten percent to move beyond their initial reading of a text," so that there is "little evidence of well-developed problem-solving strategies or critical-thinking skills."

This is probably about as far as the National Assessment ought to go in explaining itself. Setting levels of failure, mediocrity, or excellence in terms of NAEP percentages would be a serious mistake. Whatever educational standards are developed in this country will properly be made *operative* only at the local or possibly the state level. Even the making of judgments about what is "good" or "bad" is reasonable and responsible only in terms of particular educational environments.

The problem remains, if the educational standards concept is to be implemented effectively, of translating statistical gibberish into terms people can understand. Not just the National Assessment is involved. There must be some agency or forum in which the accumulating mass of education data can be brought together, synthesized, and reported so as to be generally understood.

The proposal is made in Recommendation IV that an Educational Assessment Council be established. Entirely independent of the National Assessment agency, the Council would include prestigious professional and lay members and be adequately staffed. It would not set standards. How far it would go into judgments about good and bad would depend on its consideration of all the factors involved. The Council would look carefully at the data produced by the National Assessment and take whatever steps might be necessary to inform both the general public and the teaching profession about the significance of what are in themselves meaningless numbers and percentages.

Recommendation II-G: That National Assessment results be reported on an aggregated-item basis; that recently adopted practices regarding interpretation of these results be maintained; that no definitive qualitative judgments be included in Assessment announcements; and that an independent council be established (see Recommendation IV) with the responsibility, among others, to improve public understanding of this type of data.

Research And Development

Although the early years of National Assessment history were characterized by innovation and the development of new educational measurement techniques, emphasis has shifted to the maintenance of a system already in place. Increasing costs

THE ASSESSMENT

and tightening budgets have left resources sufficient only to meet day-to-day operational demands. A good many of the proposals made in this section of the report come from NAEP officers and staff or have their full approval. The constraining factor has not been a limited purpose to improve the Assessment but rather lack of the wherewithal to develop and test out new approaches.

It was recommended earlier in the report that the Assessment not be maintained, at least under separate institutional aegis, unless the program can be restored to a cost-effective basis. This doesn't mean just making two assessments per year. Increasing the operational budget alone would be a mistake. There is equal need for a research and development component that will make the annual assessments part of an effort to improve the measuring component in establishing responsible and effective educational standards.

Although this need is clearest in connection with developing objectives and items, it pervades the entire assessment process from the initial identification of areas to be measured to the final reporting of results. The recent Reading, Thinking And Writing report repeatedly raises points that could obviously have been further explored if they had been thought of earlier in the assessment process.

It would be a mistake to attach to the Assessment a research and development unit designed to look beyond the measuring function itself. The R and D patterns worked out by the Bureau of the Census and the Bureau of Labor Statistics are helpful models. But without staff capacity to analyze both the effectiveness of current practices and the potential of alternatives, the Assessment will waste its experience and lose its promise.

Recommendation II-H: That a fuller research and development component be included in the Assessment program.

The Service Function

Implicit in Recommendation II-A is the recognition that even if the NAEP program were preserved on a one-assessment-per-year basis, it would have substantial value to state and local assessing and standard-setting agencies. If the Assessment is restored to its potential as a source of public information, the additional provision of services to these other agencies could multiply its value.

A good deal of service of this kind is already being provided, with the demand exceeding in some respects what can be met within current NAEP budgetary and staff limitations. The service pattern established so far is irregular and less efficient than a larger scale operation would be. This service potential needs to be

THE ASSESSMENT

carefully analyzed, with due recognition given its appropriate limits as well as its possibilities.

A constructive operating principle governing this service function will be drawn from identifying whatever Assessment capacities are unique. Its nationwide data base is one of these, for it opens up possibilities of use by States and local school systems seeking to compare their own situations with others. NAEP offices are also in a position to provide, as they already do to a substantial extent, full reporting of the processes followed in developing assessments.

Perfecting the National Assessment presents both the opportunity and the obligation to pioneer measuring techniques, as few other agencies can. Only as it develops clearly superior assessment elements — carefully designed spectrums of objectives, for example, and superior instrument items — will the Assessment be properly drawn on by other assessing agencies. Its public funding does not warrant its abuse as a bargain shop.

A fairly extensive program of NAEP informational publications and conferences has been developed. Some 110 documents cover the assessment reports, set out objectives and exercises, and describe methodology. A NAEP Newsletter is published quarterly. The Large Scale Assessment Conference provides an annual opportunity for communication and the exchange of ideas and experience among those interested in educational assessment. But the larger National Assessment service potential lies in the development of cooperative programs with assessment and testing offices in various States.

In 1977-78, when the Texas legislature was considering the enactment of a minimum competency testing program, the Texas Education Agency made extensive use of NAEP materials in conducting a statewide survey (Texas Assessment Project — TAP) of student achievement in Reading, Writing, Mathematics, and Citizenship. The sampling plan was patterned after the National Assessment. Both the Writing and the Citizenship assessments were based largely on items and exercises selected by a Texas Education agency staff panel from among those provided by NAEP offices.

After the Texas assessment had been completed, extensive comparisons were made between the Texas results and available NAEP data, and reported to the legislative committee for consideration in connection with the adoption of the "Texas Assessment of Basic Skills." The circumstances under which the legislation was adopted preclude any clear identification of the effect of the comparisons. There is more evidence of substantial influence of the TAP initiative on

THE ASSESSMENT

the Framework For The Social Studies and Language Arts Framework which have been developed and on the State Board Goals which have been set for 1983.

Larger potential for National Assessment usefulness is suggested by the ten years or so of cooperation between NAEP offices and the Connecticut State Board of Education, in connection with the administration of the Connecticut Assessment of Educational Progress (CAEP). A 1980 State Board report notes that "the CAEP program is modeled after the National Assessment of Educational Progress (NAEP) in its basic goals, design, and implementation." This is clearly reflected in the pattern of the twelve Connecticut assessments in seven subjects also covered by NAEP surveys.

The CAEP sampling design is like NAEP's, except that students are assessed at grade rather than age levels. Goals and objectives used for the Connecticut assessments parallel clearly the objectives and subobjectives identified for the National Assessment. Many CAEP items are NAEP items; this was true of all items in the 1979-80 Connecticut Science assessment.

The Connecticut State Board has relied extensively on comparisons of CAEP and NAEP results, even to the extent of issuing formal recommendations that special emphasis be placed on instruction in areas (for example "physical and earth science facts and principles" at the eighth and eleventh grade levels) in which the Connecticut performance appears to be below national or regional norms as indicated by NAEP results. There are repeated reflections in the record of the larger or at least easier use which could be made of the NAEP data if it were developed on a grade-level rather than an age-level basis.

The Connecticut experience includes the use of assessment results to inform local school system administrators, curriculum officers, and teachers regarding various aspects of their functions. The recently established Connecticut Comprehensive Plan for Elementary and Secondary Education carries these uses to a new level of sophistication.

THE ASSESSMENT

The plan calls for a regular cycle of assessments in various subjects at five-year intervals, maintenance of comparable objectives and items to permit comparisons over time, and the inclusion of enough NAEP items to permit comparisons of performance with nationwide and regional levels.

Comparable uses of National Assessment materials have been made in a number of other States. A recent NAEP staff summary lists twelve States as having closely replicated the National Assessment model, fourteen as having used NAEP materials to complement their own assessment model, and twelve others as having drawn on NAEP offices for technical and consultative advice. There is clear confirmation in this record of not only a substantial service potential but also of a significant prospect for integrating state and nationwide assessment programs.

Recommendation II-I: That the Assessment program be designed and administered to maximize its service function to state and local educational assessment and standard-setting agencies.

Summary

Despite its values, maintaining the National Assessment at the current level of operations in a starved condition would be a mistake, for those values are being neglected. Yet losing them would be a worse waste. The revitalization of the Assessment program demands taking a number of fairly specific measures.

The current budget means one nationwide assessment per year. The program must be restored to permit two assessments each year. (Recommendation II-A.)

Although enlarging the present NAEP sample would increase its usefulness, cost considerations

THE ASSESSMENT

preclude this. Alternative possibilities should be considered, however, to permit using the Assessment results to provide state-by-state comparisons where this is desired by state officials. (Recommendations II-B and C)

With new emphasis on establishing educational standards by most States and many local school systems, and in view of the structuring of state and local systems on a grade-level basis, the National Assessment policy of taking measurements at age-levels should be revised to permit accommodation to the general practice. (Recommendations II-D)

The nerve center of any educational assessment or standard-setting system is in the process by which objectives and instrument items are developed. The key demand on the National Assessment is that it assume leadership in the development of objectives and items that not only permit comparisons but also contribute to improving the quality of education. This involves process and techniques. It also goes to underlying principles. (Recommendations II-E and F.)

At least until recently, Assessment results have been ineffectively reported. The traditional practice of item-by-item reporting and abstention from interpretation vitiated much of its potential value. Recent changes, including aggregations of item results and the issuance of interpretative statements, should be carried forward. But the ultimate conclusions as to the levels of student achievement that are to be considered good or bad must be left to the users of the Assessment information. (Recommendation II-G.)

Fulfillment of the Assessment's purpose and potential requires that the research and development component of its program be enlarged. (Recommendation II-H.)

A revitalized National Assessment program will include providing expanded service functions to state and local assessment and standard-setting agencies. (Recommendation II-I.)

PART III

Assessment,
Analysis,
And
Policy
Development

POLICY DEVELOPMENT

The recommendation to maintain and strengthen the traditional functions of the National Assessment does not directly reply to its sharpest and most constructive critics. They assume or at least accept the notion that measuring student achievement is valuable. They criticize the failure to develop the full potential of the National Assessment as an instrument of educational policymaking.

Two distinct issues arise here. The first involves the relationship between the assessment function and the analysis of the information produced. The administration of the National Assessment is criticized for having frustrated analysis which could make the data much more useful.

It is also complained that the effectiveness of the data is diminished because they are not focused on particular educational policy issues. The point is pressed that with certain educational problem areas having been identified, the Assessment should be so structured that the data produced shed light on these problems.

The recommendations in Parts I and II of this report are, in net effect, that the central character of the Assessment — as a statistically reliable measurement of the nationwide results of whatever is affecting the educational process — be maintained. The considerations involved here go beyond statistical perfectionism. One of the major values of the Assessment is that in refining the measurement of student achievements — by a thorough analysis of educational objectives — it can contribute uniquely to the development of a quality concept of educational standards. In this important respect, the Assessment goes beyond measurement to become a critically important instrument of educational policy development.

Considerable justification nevertheless remains for the criticism that the Assessment has not been put to its fullest possible use in influencing educational policy and decisions. Meeting these criticisms does not require compromising the Assessment's essential principles or diluting its central values. It calls rather for measures to facilitate drawing from the Assessment data information which is already there, and for some extensions of the assessment process.

Assessment And Analysis

It is oversimplifying to define educational assessment as the measurement of educational outcomes, and educational analysis

POLICY DEVELOPMENT 41

or research as the identification of the causes of these results. There is no such clear cut line. It is also a misconception that the relationship between these two functions, so far as the National Assessment is concerned, turns entirely or even primarily on the balance between the number of "achievement" items and "student background" items included in a particular assessment instrument. What is involved here goes substantially beyond this.

National Assessment practice has been to identify seven student variable items: age, modal grade, race, gender, parental education, type and size of community, and section of the country. A good deal of additional background information is obtained. The 1975-76 Citizenship and Social Studies Assessment, included, for the seventeen-year-old group, a supplemental questionnaire covering twenty types of student background and educational context information. (The regular assessment packages were not shortened to allow for this; instead a ten-minute period was added to the hour the students spent on the packages.) The supplemental questionnaire covered a variety of matters: socioeconomic status, study habits, television watching, language spoken at home, ethnic heritage, number of siblings, educational and career expectations, instructional experience, extracurricular activities, self-conceptualization, attitudes toward education, and so forth.

Subsequent assessments have included abbreviated forms of the supplemental questionnaire in the regular one-hour package. The 1979-80 Reading/Literature Assessment instrument for seventeen-year-olds contained in each package ten items covering different kinds of information about students' circumstances.

The difficulty here traces less to failure to collect this type of information than to problems in making effective use of it. The regular NAEP reports permit attempts to correlate achievement levels with the seven core variables but the other supplemental information is stored on the Assessment computer tapes. Until recently, researchers and analysts have had little effective access to these data.

This problem is not limited to analyses of the student background information. The NAEP data-base would permit constructive research into many other areas: interdisciplinary achievements, relationships between content and process responses, attitudinal elements, teaching and testing methodologies, and so forth. But little of this has been possible as a practical matter.

This difficulty has been a major issue of NAEP administration. Starting in 1974, a limited number of data files were developed and disseminated to a few researchers. But the files

POLICY DEVELOPMENT

covered only a little of the accumulated Assessment data and were in a form making their use exceedingly difficult. There were strong protests about this.

During the past two years, principally at the insistence of National Institute of Education (NIE) officers, significant efforts have been made to improve the situation. With grants from NIE, the National Science Foundation, and the Education Commission of the States, NAEP staff began developing "Public Use Data Tapes" to permit easier use by analysts. Although only a portion of what is in the NAEP data-base is yet on these tapes, NAEP staff activity is being concentrated increasingly on putting the accumulated data into readily accessible shape.

A series of research projects to develop access techniques has been instituted. The NAEP matrix sampling process, which involves dividing items into different packages given to different students, seriously complicates the correlation of achievement item responses with student background information. Complex processes of adapting packages for computer programming (such as SPSS — Statistical Package for the Social Sciences; or SAS — Statistical Analysis System) are involved. Research analysts have to be trained to use these computer programs. Catching up on the NAEP backlog is proving a formidable task.

The view is strongly held by some that the purpose of the National Assessment could be adequately served by simpler sampling designs posing greatly reduced data retrieval difficulties. The familiar charge is that NAEP architects concentrated too much on creating a "statistical showpiece" and too little on producing usable data. Statisticians respond that the NAEP design is necessary to assure the validity of the student achievement information. The debate involves elements which defy responsible resolution by non-experts.

But it develops that the NAEP data *can* be put in form accessible and meaningful to analysts, and that once the system for doing this is in place, the costs of maintaining it appear manageable. Under these circumstances, the traditional debate — about whether the National Assessment is to be maintained as a statistically reliable descriptive measurement of educational results or whether it is to be adjusted to meet the interests and demands of research inquiry into the causes of the results disclosed — emerges as essentially false.

Experience confirms the capacity of the Assessment to accommodate in its instruments, without impairing the statistical integrity of the measurements, a substantial number of the types of items considered important in

POLICY DEVELOPMENT

analyzing various hypotheses about the causes of these results. Perhaps improved procedures should be developed for deciding which student background or educational context items will be most helpfully included in the regular National Assessments. What is important is that the incomparably broad and valuable Assessment data-base be made readily accessible to those who are analyzing the information about what is happening in education in order to determine the reasons for these developments — as an essential step toward deciding what to do about the situation. This is critical to the realization of the National Assessment's full potential as an instrument for education's improvement.

Recommendation III-A: That Assessment data be developed in forms facilitating their use for research purposes, including particularly the analysis of factors that may relate causally to student achivement.

Specific Assessments

Parts I and II of this report recognize and confirm the importance of maintaining a reliable nationwide index of students' educational achievement. But the value of this kind of information would be enhanced if it could also be focussed on particular issues of special interest to educational administrators, teachers, and the general public. The National Assessment provides a base for developing a program of "specific assessments." Several examples will suggest the possibilities.

One. The equity/quality equation continues to haunt the country. The present NAEP data-base includes much relevant information. The responses of more than 900,000 students on achievement items covering ten subjects can be analyzed along with the identification of their race and their parents' education. These data permit identifying changes during a decade of effort to undo the damage of racial discrimination.

Increasing the accessibility of this data-base will unlock this accumulated information. But its value would be multiplied by designing a specific assessment to cover this particular problem area. More background items could be used, perhaps including some from the 1975-76 Citizenship and Social Studies Assessment. Key achievement items could be taken from previous assessments in several learning areas. The sample for the specific assessment would be developed to meet its particular purpose.

Two. Current interest in bilingual education would warrant constructing a similar specific assessment. The NAEP achievement data developed since 1975

POLICY DEVELOPMENT

can all be related to students' identification of their Hispanic origins and to language spoken in their homes. The research potential of these data alone is significant but limited. Invaluable information could be assembled by using these data as a base for constructing a specific assessment concentrated by item selection and sampling design on the bilingual education issue.

Three. Various questions are developing about the effectiveness and the possible side effects of minimum competency testing systems. The NAEP data have been collected during the period when these systems were being introduced. Although these data don't permit reliable state-by-state achievement level assessments, each student's state of residence is tabulated — which may (or may not) provide a comparison basis. In any event, a specific assessment could be constructed to illuminate, over time, what happens to achievement levels — not only at the bottom of the scale but also nearer the top — in States that do, and others that do not, concentrate on minimum competencies. Such an assessment would at least provide a basis for more refined research and analysis.

Four. A different kind of specific assessment would help correct the tilt in the educational standards concept toward functional literacy and away from excellence. The Commission on Educational Excellence, established recently by the Secretary of Education, could be asked to identify schools generally recognized as superior, taking account of both the problems they face and the product turned out. Arrangements could then be made to administer the regular National Assessment instruments to a sample of students from these schools.

This specific assessment would have the value of providing at least a clue to what could properly be considered "good" on the scale suggested by the National Assessment results. States or local school systems using NAEP materials could make comparative evaluations of their own assessment results. There could be no suspicion of this as federal standard setting. The plainest value would be the reminder that educational excellence is important.

Five. One or more specific assessments could explore the possibilities of illuminating student achievement by identifying more of its educational context. This is a different matter from including "student background variables." Getting into the individual students' personal characteristics and situations invariably prompts warnings that the NAEP purpose is not to analyze human development, and injunctions against confusing the measurement of educational results (outcomes) and the analysis of causes (inputs). But it is being recognized increasingly that the measuring of achievement is incomplete without an accompanying identification of whatever educational circumstances may affect these results.

This has been partially recognized from the beginning by the NAEP identification of the types of schools students attend. National Assessment data can be analyzed — as was done earlier this year — in terms of whether the schools are supported publicly or privately. The number of courses a student has had in the subject being assessed is also determined.

The Citizenship and Social Studies Assessment, the 1976-77 Science Assessment, and the 1977-78 Mathematics Assessment all included questions about the teaching methods to which the students had been exposed. Questions about homework have been added several times. The Writing assessments provide quite a bit of teaching methodology information. Several of the 1981 NIE research projects are designed to draw educational process implications from the achievement items in the 1977-78 Mathematics Assessment.

These initiatives should be pulled together. Various research studies confirm the possibility of isolating educational context elements for analysis: "time on task," the kinds of textbooks being used, curricular patterns being followed, extracurricular activities, guidance and career counseling provided, perhaps (though this gets into a sharply controversial area) something about the teachers' training. The measurement of student achievement is being increasingly recognized as incomplete except as it provides insight into the schooling process. Specific assessments specially constructed to illuminate this area would be invaluable to administrators, teachers, and researchers. These are things parents want to know — and should.

Six. In some people's view, the most meaningful and significant measuring of education's effectiveness and progress requires following particular groups of students over periods of time, reexamining them and determining how they make out. Most of the advice received in connection with the preparation of this report is that it would be a mistake to try to use the National Assessment for such "longitudinal" analysis. Yet the argument supporting such measurements is persuasive. A program of specific assessments could include at least exploration of this longitudinal study potential.

If assessments in the basic subject areas were made every four years, achievement levels of a characteristic cross section of the same group of students could be compared in the fourth, eighth and twelfth grades. Perhaps there is larger promise in trying to develop a specific assessment, adjunct to a general assessment, which could be related in some way to one of the other longitudinal study systems. An attempt to establish such a linkage resulted in adding the Supplementary Student Questionnaire to the 1975-76 Citizenship and Social Studies Assessment. Although nothing came of that effort, this possibility warrants further exploration.

Seven. Other specific assessments might concentrate on the measuring process itself. Attempts to measure the interrelated effects on student achievement of formal education and the instruction of broader experience continue to feed rather than reduce controversy. In an entirely different connection, item response theory is

POLICY DEVELOPMENT

receiving increasing attention, and two of the 1981 NIE research projects are inquiring into the relevance of this theory to NAEP data. Perhaps such matters are better left for the present in the research domain. But an experimental specific assessment might at least test the possibilities of providing a basis for this research.

Some will warn against diluting the National Assessment's central purpose by trying to make it too many things to too many people. The other possibility is that too few people will find broad nationwide student achievement level data sufficiently valuable *except* as their illumination can be focused on issues of generally recognized importance. "We need a national assessment...to help the dialogue reach occasional conclusions so that new dialogues may spring up and progress" (Timpane, 1981).

Recommendation III-B: That a program of specific assessments be developed to illuminate particular educational policy issues.

Summary

National educational assessment, obviously not an end in itself, is worthwhile only as it contributes to the formulation of more effective educational policy. Parts I and II of this report affirm the value to this policymaking process of maintaining and strengthening the present National Assessment as a reliable and broad measure of the overall educational condition.

The worth of the assessment function will be increased, however, if the information it provides can be used in analyzing the causes of educational results and in considering possible remedies for whatever malfunctions are disclosed. Recommendation III-A is that procedures be developed to facilitate the analysis of NAEP data through secondary research and analysis which will illuminate their implications. This recommendation is directed primarily toward administrative practices that have in the past taken too little account of these essential functions.

Recommendation III-B is that a program of "specific assessments" be developed to illuminate particular educational policy issues such as the twin demands for educational quality and equity, bilingual education, the effects and possible side effects of minimum competency examinations, and the efficacy of various teaching and educational measuring techniques. These more sharply focused assessments will supplement and complement traditional National Assessment measurements.

PART IV

The Overview Function

OVERVIEW FUNCTION

The danger in looking closely and intently at the National Assessment is that an inclination develops to try to put too much load on this one part of an obviously wider system. The Assessment program should be improved and strengthened. But the largest promise of increasing its effectiveness lies in bringing more order into the educational assessment system as a whole.

This system has become a formidable complex of public and private, local and state and national, cross-sectional and longitudinal measurings of general and specific achievements, aptitudes, and abilities. The teacher, the school system, the State, the college entrance examiners, and the NAEP assessor all ask a student how well she or he can read. The various collections of data are distinguishable but they unquestionably overlap. Much of what is found out is put to only a fraction of its possible use. While some parts of the picture are over-exposed others get no light at all. The clear value of trial and error gets lost because it isn't communicated. Having grown rapidly, the educational assessment system is healthy, promising, and so awkward that it falls over itself.

Although this situation suggests neither crisis nor emergency, two elements clearly warrant attention and appear in no way formidable. The possibilities of first, fuller development, and second, better coordination of the overall education assessment system are obvious. It is equally clear that the lines of communication between the educational measurers and the users of their measurements, especially the general public, need — and permit — substantial improvement.

Any suggestion of introducing centralized control elements into this situation would be immediately and properly rebuked. No significant competing interests are involved. What is called for is a forum and process that will permit and prompt those who are actively interested to take effective common counsel.

Going perhaps beyond the charter of this report, it is proposed that an Educational Assessment Council be established. Risking disagreement about particulars in order to make the suggestion clear, the proposal is made in specific form.

The Educational Assessment Council (EAC, or just the Council) would include six to eight members. Some of them would be educators, some of them not. They would be prestigious and distinguished men and women who would commit from fifteen to twenty-five days of time each year to the Council. Its staff would include three or four full-time, professionally trained persons.

OVERVIEW FUNCTION

Funding and locus are obviously important details. This report assumes a sufficient continuing Congressional interest that the legislation authorizing the National Assessment will be renewed in 1983. Provision for an Educational Assessment Council could be made in the same statute. The appropriate federal agency would be authorized to make the necessary contract or grant arrangements with a private institution, which might or might not be the institution administering the National Assessment. The Council would be an entirely separate unit, autonomous and independent in every respect affecting policy and decision making.

If federal support is not forthcoming, effectuation of the Council proposal will probably depend on private foundation interest. The contemplated function would be within the ambit of various institutions and organizations, universities, research centers, and educational associations — perhaps particularly the National Academy of Education.

The Council would address its attention particularly to the two needs which have been mentioned. The communication problem might be given priority because of its relative immediacy, with the attempted rationalization of the assessment system being approached more slowly.

Better Uses Of Available Information

The proposed Council, working with its staff, would stay abreast of all educational assessments being made and of related testing and measuring and evaluative programs and studies. This would include the National Assessment reports, the specific assessments proposed in Recommendation III-B, and the state and local assessment programs. Recognizing the different purpose and emphasis in minimum competency testing programs, the Council would bring them into the same picture. Reports on the various longitudinal studies being made would be followed closely, and the invaluable data being gathered by the National Center for Education Statistics (NCES) would be factored in. Close attention would be given evaluative research inquiries into closely related matters.

The Council would look into possible ways of synthesizing and interpreting the information provided by current educational measurements. Considerable amounts of data emerge each year but are never put in a form permitting their maximum use, especially by the public, but also by educational administrators and teachers. A system of regular

OVERVIEW FUNCTION

critical reporting on this variety of information, by professionally competent and prestigious persons, would be infinitely valuable to educators and the general public alike.

The Council would give particular consideration to the possibility of occasional *net* appraisals of "educational progress" and the "condition of American education" — which both the National Assessment and the NCES reports claim to be by their titles but cannot possibly be. The value and influence of such a net appraisal would derive both from the distinction of the Council membership and from the unique completeness of the review.

The Council would determine the form in which such an overall report would be most effective and to whom it should be addressed. There would unquestionably be widespread public interest in the appraisal. Such an appraisal, as well as the Council's periodic reportings, would have obvious value to chief state school officers, to urban and rural superintendents and principals. Teachers would be vitally interested. Transmittal of the report to the Congress, the Secretary of Education, and perhaps to the President would be considered, with account taken both of the functional justification for doing so and of any appearance that such action would disregard sensitivities about the federal role in education.

Repeated reference has been made in this report to the problems which inhere in reducing students' academic achievements to statistics and then trying to communicate the meaning of these numbers to those who are concerned — the students themselves, parents, analysts, educational policy makers, and the general public. Giving new significance to educational standards makes this problem more serious. The whole notion of public accountability depends directly on these statistics being reliable to begin with and then on their interpretation with reasonable accuracy and understanding.

As the situation stands today, a jumble of figures on educational achievement is constantly emerging. Conscientious parents might see their own child's report card (perhaps in computer print-out form), read the newspaper reports on the results of a local or state minimum competency examination, look at media stories about the college entrance examinations, see other reports on the National Assessment results — and shake their heads in complete confusion. There is a good deal of illusion in the notion that people in general know from these figures very much about what is actually happening in education — or what to do about it. The fault is not in the figures but in a failure of communication.

OVERVIEW FUNCTION

Something like the proposed Educational Assessment Council could contribute substantially to improving the critical understanding. In interpreting the bewildering data, the Council members would inevitably be making value judgments. They would be properly careful about this. But the data themselves are based on value judgments, made as part of the measuring process. Other value judgments control what emphasis is placed by the media on reporting these statistics. The Council's reports would be fully informed, with the benefit of competent staff advice. The reports would be responsible by virtue of the Council members' calibre. Whatever risk of value judgments is left is worth taking. It is probably essential to making sense out of the idea of educational standards and public accountability.

An Improved Assessment System

This report has emphasized the difference between the function performed by the National Assessment and the broader reaches of educational assessment. The report deals to a limited extent with the possibilities of fuller coordination of the National Assessment with state and local systems. It also suggests adding certain specific assessment functions to the traditional NAEP program. Limited aspects of the relationship between the assessment and analysis functions have been covered.

This leaves a broad area of much needed overall system analysis. The proposed Educational Assessment Council would properly undertake this analysis.

The Council would give concentrated attention to identifying currently unmet assessment needs. This would involve special consideration of how the system can be developed to meet more fully and directly particular concerns of educational administrators and the interests of the educational research community.

Part of what is contemplated here was suggested in connection with Recommendation III-B. Although operational responsibility for most of the specific assessments proposed there would lie with the unit administering the Assessment, the Council would be in a position to exercise broad influence on the kind of specific assessments that would be undertaken.

It is not assumed that the Council would itself undertake any assessment or other measurement function or engage in extensive research. It would have limited funds to commission research projects. Its prestige, competence, and broad perspective would give it leverage in the development of a more complete and rational architecture for existing educational assessment and measurement systems.

OVERVIEW FUNCTION 52

A small Council and staff will obviously not be a control tower. But somebody ought to be working out and suggesting possible flight patterns.

To concentrate, as this report has, on one part of the educational assessment system, is to be properly concerned about myopia. Trying to figure out how a system in place can be made to work better, relying on the principle of incremental policy development, risks thinking too narrowly. Somebody ought to be asking whether there may be as much fundamental truth still undiscovered about testing and educational measurement as there was until recently about the mutability of genes and the force inside atoms.

An Educational Assessment Council would not set up its own laboratory. It would find out whatever is currently suspected or responsibly conjectured about educational measurement's frontiers and horizons, and it would press others to step up their exploring.

The Council would probably consider, for example, what could be advisedly suggested to the broader assessment community regarding item response theory. Somebody should be asking whether the entire educational assessment procedure should be rebuilt around different principles.

Experience with social indicators, in education and in other fields, is accumulating in several countries. The absence of more than passing reference to their experience is a recognized weakness of this report. A survey of what has been done in Western Europe, Scandinavia, Great Britain, and Japan, and by the Office of Economic and Community Development would be high on the proposed Council's agenda.

Recommendation IV: That an Educational Assessment Council be established to synthesize data developed by various assessment and measuring systems, to improve communication of the meaning and significance of educational statistics, and to recommend changes in the processes and structure of the educational measurement system.

In a recent volume, Toward Reform Of Program Evaluation, Lee J. Cronbach and his associates analyze the program evaluation discipline. They conclude their "social problem study groups" be set up to perform, in connection with program evaluation, functions much like those outlined here for an Educational Assessment Council. The point is made that "a professional, sophisticated evaluation community, free to do its best work, will educate all sectors of the political community. Out of that understanding will come more realistic demands for social services, more stable policies, and services that merit the trust placed in them." Recognizing assessment and program evaluation as separate and distinguishable functions, this still makes common good sense.

PART V

Implementation

IMPLEMENTATION

Recommendations about implementing the program proposals made here are complicated by what seems almost a conspiracy of current uncertainties. With the present authorization expiring in December 1983, the Assessment program will come before the Congress for review and reappraisal, leading to renewal or amendment or elimination, during a period when the entire federal role in education is under reconsideration. If Congress acts affirmatively, responsibility for carrying out its action will then be placed in an as yet unidentifiable unit somewhere within a federal executive structure currently facing major reorganization. There will be a selection, in a year or two, of the private organization or institution that will administer, on either a grant or a contract basis, the newly authorized national educational assessment program. It would be presumptuous and improper to conjecture about what private agency this will be.

Although it is tempting under these circumstances to leave the whole matter of implementation to the political fates, that much timidity would mock the importance of the present inquiry. If this report is to inform the political process which will determine the Assessment's future, the unusual fluidity of the process must be viewed not only as a fact to be allowed for but as a potentially advantageous force. The situation commends, instead of specific recommendations, pointing out alternative ways to implement the principles underlying the program proposals.

The critical determination is for the Congress to make, next year or in 1983, in renewing or changing or possibly terminating the NAEP authorizing legislation. The recommendation in this report is that the Assessment program be continued in a modified form.

If the Congress decides to continue an educational assessment program, it will have two alternatives. One would be to leave the Assessment authorization about as is, simply extending Public Law 95-561, which authorizes the present program, for an additional period of time, adjusting the statute to whatever structural changes may be made in what is presently the Department of Education. This would involve minimum effort and would have the support of most of those who have been closest to this program.

IMPLEMENTATION

That action would accommodate the proposals made in Parts I, II, and III in this report. Administrative action necessary to implement these program proposals could be taken by the federal executive office made responsible for this legislation and by the private institution or organization selected to administer the program.

An alternative course merits at least equal consideration. It would take into account the possibly adverse reactions, in the present political context, to the Assessment's history. It was instituted in the mid-1960s at a time when different views prevailed regarding the federal role in education. It has not, by objective measure, established its reputation firmly during twelve to fifteen years of operation. Accepting the importance of the assessment function, it may well be considered appropriate to mark clearly, by programmatic and institutional changes, the entering of a new phase of educational measurement related to the new concept of educational standards.

Though the program recommendations made in this report draw substantially on NAEP practice and experience, they represent major changes in emphasis and purpose. It would be a reasonable conclusion that these changes, and those that others may propose, could be more fully and effectively implemented as part of a clearly identifiable new initiative, including a new administrative procedure and organization.

Action along this line might include the setting up of an "Educational Assessment Center." One unit of the Center would be responsible for administering the traditional Assessment and probably the "specific assessments" proposed in Recommendation III-B. This unit might or might not include the research and development function suggested in Recommendation II-H. The Council proposed in Recommendation IV would probably be established within the Center organizational structure, but on an entirely autonomous basis. If this appeared difficult to accomplish, it could be set up as an entirely separate entity.

IMPLEMENTATION

It would be appropriate in a revised Congressional authorization to emphasize the assessment program's relationship to educational standards, the importance of its service relationship to state agencies and local school systems, and the imperative of leaving to those systems the determination of performance levels that are to be considered satisfactory.

Responsibility for implementing the Congressional authorization was placed earlier with the National Center for Education Statistics, and is currently in the National Institute of Education. Both agencies are in the Department of Education. The current review of that structure makes it pointless to conjecture or recommend regarding the placement of responsibility for administering whatever assessment program may be authorized.

The sound suggestion has been made that the contract or grant arrangements entered into with a private organization be made for an eight to ten year period rather than for four years, the present practice. It is important that the contract or grant mechanism assure completely open competition for the contract or grant.

The present statute provides simply that NIE make arrangements to "carry out (the National Assessment program) by grant to or cooperative agreement...with a non-profit education organization."

For the past twelve years, the Assessment has been administered by the Education Commission of the States (ECS). The most recent grant by NIE to ECS was made in 1979, covering the period through 1983.

In practical organizational and operational terms, the Assessment program is administered through a virtually autonomous unit. The statute places broad responsibility on a seventeen member Assessment Policy Committee (APC), and prescribes the composition of the Committee: three public members, two from business and industry, one chief state school officer, two State legislators, two school district superintendents, one chairman of a state board of

IMPLEMENTATION

education, one chairman of a local school board, one Governor of a State, and four classroom teachers. The seventeen members are appointed from among these categories by the administering organization. The Director of NIE, an ex officio member, also appoints a member of the National Council on Education Research as a nonvoting member of the APC.

One element emerges from experience as warranting special consideration in the future. As an almost inevitable consequence of the statutory prescription for a large, representative governing body, an unusual degree of responsibility has devolved on the chief executive officer. The Director is selected by the Assessment Policy Committee and reports to it. The APC determination on all policy issues is controlling.

The APC meets only three times a year, for two-day sessions. The agenda is invariably long, including myriad items covered for purely formal reasons. There is usually very little time for policy making. The fact that so many APC members are institutional representatives complicates getting beyond superficial and sometimes enervating compromises. Complete respect for the incumbent membership of the APC and for NAEP executives is no reason to soften the conclusion that this arrangement has resulted in less than adequate exercise of affirmative and constructive policy development responsibility.

This problem will become even more acute if the Assessment program is developed along the lines proposed in this report. A large committee, including representatives of various NAEP user groups, can contribute a great deal but should probably serve an advisory function. Effective administration of the program envisaged here will require that policy responsibility be exercised by a board of directors constituted to function in the usual private corporation pattern.

IMPLEMENTATION

Summary

It is appropriate to indicate the cost implications of the various proposals that have been made here.

Continuing the present NAEP program at its present levels now reduced to roughly one Assessment completion each year, will cost between $3.5 million and $4 million per year. Appropriations have recently been worked out for 1982 operations at an annual figure of approximately $3.9 million. It was suggested earlier in the report that proceeding on this basis leaves serious questions as to whether the point of marginal utility has already been passed. The more enduring consideration is how worthwhile and how cost-effective a broadened and strengthened Assessment program would be.

Programming two assessments each year, as suggested in Parts I and III, would cost about $2 million more per year.

Although the research and development unit proposed in Part II could be set up on various possible scales, a cost estimate of $500,000 per year appears appropriate.

The Educational Assessment Council function (Part IV) is critical. Salary and related costs for the Council itself and for a small staff would amount to between $300,000 and $400,000 a year. Enabling the Council to authorize and fund a variety of research and evaluation projects will enlarge its capacity and will result in significant contributions to the educational measurement and standards-setting processes. Careful consideration leads to the recommendation that $1 million be authorized and appropriated for the Educational Assessment Council.

This adds up to a recommended total budget of $7.5 million plus whatever costs of federal agency administration are involved. This compares with a current NAEP expenditure level of $3.9 million. The annual authorization ceiling in the present statute is $10.5 million.

One other comparison is appropriate. Current annual expenditures in this country for educational assessments and tests of one kind or another are estimated at between $150 million and $200 million. It is a conservative judgment that the effective administration of a carefully constructed Assessment program will permit economies throughout the existing educational measurement system which will much more than pay for the program's costs.

CONCLUSION AND SUMMARY

CONCLUSION

This report on the National Assessment of Educational Progress is appropriately summarized in terms of immediate political realities. The Congressional authorization of the Assessment expires at the end of 1983. The last two assessments under the existing program are now in progress. The operative question is whether the Congress will, some time within the next twelve to eighteen months, extend the present NAEP authorization, terminate the program, or re-establish it on modified terms.

The basic recommendation made here in Part I, is that the National Assessment program be maintained and materially strengthened. This is for one central reason. During the past few years, a decision has been made in this country to rely for the improvement of education on the establishment of higher educational standards and on a principle of public accountability. The National Assessment has a unique capacity for implementing this decision.

The distinguishing characteristic of the Assessment is that it provides a measurement of the academic achievement levels of students throughout the entire country. The process relied on is a model of statistical integrity and accuracy. Measurements have been taken at four age levels: nine, thirteen, seventeen, and twenty-six to thirty-five. Assessments have been made in various subjects at four-year or five-year intervals. The techniques which are used permit reliable determination of changes in educational achievement levels over time.

The National Assessment serves three sets of functions. Its reports inform the general public and educational policy makers about the changes taking place in student achievement levels. Its internal processes provide a superior opportunity for developing the critical elements of standards that will contribute to the improvement of education. Not itself a standard-setting agency, it provides invaluable services to state and local educational assessment and testing and standard-setting offices.

The recommendation is not that the Assessment program be continued at the present operational level. It should not be. Funding for the Assessment has been reduced from $7 million in 1973 to a current level of less than $4 million. Cutting the program back has made it ordinary, diminished its vitality, and left it unable to play the role it should in implementing the educational standards and public accountability principles. The critical recommendations in this report are therefore those contained in Parts II, III, and IV — for strengthening and extending the National Assessment program and rationalizing the broader educational measurement system.

Part II covers the traditional function of the Assessment, as a

CONCLUSION

measurement of nationwide student achievement levels. The recommendations concentrate on developing the National Assessment as an implementing element in an educational standards and public accountability system which recognizes state and local decision and action as the only operative force. Various issues are considered in a pattern corresponding roughly with the sequence of steps involved in making an assessment — from the selection of subjects for assessment to the ultimate reporting of results.

Once two National Assessments were completed and reported each year; there is now sufficient funding for only one — and not quite enough for that. The Assessment program has covered ten subjects in various years. It has now been cut back to little more than Reading, Writing, and Mathematics. The recommendation (II-A) that the program be restored to a level permitting two assessments each year is based on the conviction that an exclusive emphasis on these three subjects would contribute to narrowing American education.

The Assessment's sampling design and instrument administration practices have been carried to extremes of carefulness. This is expensive but critical. A number of operational economies have already been effectuated. A review of this situation leads to the conclusion that further cutbacks would threaten the integrity of the measurement. The only recommndation (II-B) for change is that the one-hour limitation on the use of student time be increased to two hours.

The size of the Assessment sample does not permit state-by-state reporting of the results, and enlarging it sufficiently to do so would be prohibitively expensive. But alternative arrangements would enable state or local school officials to use NAEP results for comparisons. Recommendation II-C is that steps be taken to facilitate such arrangements.

National Assessment measurements are made at age levels — of students who are nine, thirteen, seventeen, and of young adults aged twenty-six to thirty-five. An out-of-school group has been included in the seventeen-year-old measurements. Budgetary contraints have required eliminating measurements of the out-of-school seventeen-year-old and the young adult groups. It is recommended (II-D) that funding be provided to permit including these two groups. The recommendation is also made that the Assessment be shifted to a grade-level basis (fourth, eighth, and twelfth) to bring it more into line with state and local assessment practices.

The key elements of the Assessment program involve what are in a sense the technicalities of "identifying objectives" and "selecting exercises" in constructing the Assessment instruments. But these technicalities become the heart of the standards-setting process. The development of

CONCLUSION

"higher educational standards," using this term in its fullest and most constructive sense, depends on identifying objectives that are consistent with education's ultimate purposes and most effective processes.

Recommendation II-E is that additional steps be taken to assure that NAEP objectives consist with a quality concept, rather than a comparison concept, of educational standards. Instead of determining "what is being taught" and basing the objectives on present practice, the controlling question is "what ought to be taught." The most recent National Assessment reports indicate significant movement in this direction.

The recommendation (II-F) regarding the development of exercises or items for inclusion in the assessment instrument is that a new capacity be built for developing instrument exercises consistent with the demands of a quality concept of educational standards.

The Assessment has had serious difficulty in communicating the results of its measurements effectively. This has been primarily a consequence of reporting results on an item-by-item basis. This practice has gradually been changed. The item statistics are now aggregated. Some additional interpretive steps have been taken. Recommendation II-G is that the new program be continued. But it is specifically recommended that caution be exercised against putting the Assessment results in a form that could be misconstrued as constituting national — or "federal" — standards.

Recommendation II-H is that the Assessment's internal research and development capacity be expanded.

The Assessment's valuable service to state assessment and testing agencies has received too little notice. With almost all States and many local school systems now developing their own systems, this service function becomes increasingly important. Recommendation II-G in essence simply emphasizes the importance of this service function.

The National Assessment is worthwhile only to the extent that it affects educational policy development. It does this by reporting overall student achievement levels in the country as a whole, and by developing objectives and items that add to the sophistication and quality of educational standards. But this leaves two other avenues which the Assessment administrators have not developed sufficiently.

CONCLUSION

The National Assessment reports the results — the outcomes — of the educational process. But the causes of these results, and ultimately the courses of action that can correct whatever malfunctions are disclosed, are more important.

From the beginning, the Assessment has been criticized for measuring and reporting student achievement in a form which frustrates analysts' and researchers' attempt to identify factors contributing to these results. Some of this difficulty inheres in the matrix sampling technique used in making the Assessment. There has also been continual argument about the reluctance to include as many student background items in the assessment packages as analysts and researchers consider essential for their purposes.

Part of this problem has been that the collected data have not been put in a form making them readily accessible to analysts. Computer storage and retrieval techniques have been vastly improved during the past ten years, and procedures have now been developed to meet this difficulty. Recommendation III-A reflects the conclusion that opening up the Assessment data base by using these procedures will go a long way toward eliminating an unfortunate waste of Assessment potential.

Recommendation III-B is that a program of "specific assessments" directed at particular educational policy issues be developed as an adjunct to the traditional National Assessment program. Such specific assessments might be directed, for example, at the issues of educational quality and equity, bilingual education, the effects and possible side effects of emphasizing minimum competencies, the promotion of educational excellence, and the efficacy of various teaching and educational measuring techniques. The National Assessment's measurements of overall educational achievement could be given increased effectiveness by a complementary focussing of such information on issues of particular moment.

Recognizing the National Assessment as only one key part of a much broader educational assessment and measuring system, it is suggested in Part IV that an Educational Assessment Council be created. The Council would be separate from the unit administering the National Assessment. It would be composed of educational professionals and others of established distinction.

CONCLUSION

The Council, working with a small staff, would direct its attention to two areas. One would be the communication problem which develops inevitably in trying to reduce educational achievement to numbers and then trying to make the numbers meaningful to those who rely on them. The Council would synthesize the present bewildering jumble of statistical and analytical reports, identifying their importance, and perhaps issuing an occasional overall summary of the condition and progress of education as reflected in these data and analyses. The Council would also review and appraise the entire educational assessment system and would suggest possible improvements in the present duplicative and yet still incomplete structure.

Current uncertainties preclude specific recommendations regarding the implementation of these proposals. The legislation authorizing the National Assessment program expires in 1983 and the structure of the federal Department of Education is under scrutiny. So alternative possibilities are noted: that the Congress might renew the legislation now in effect; or that a new Educational Assessment Center might be developed in a form responsive to the emergence of a new educational standards concept.

A preliminary costing out of the program proposed in this report indicates an expense of between $7 million and $8.5 million per year. This compares with the $10 million authorization figure in the current legislation, the $3.9 million National Assessment budget for the current fiscal year — and an estimated overall annual expenditure in this country of between $150 million and $200 million on educational measurement. It is implicit in the report that an expenditure in the suggested range will represent a sound investment in monitoring nationwide student achievement and in putting the idea of higher educational standards to work.

APPENDIX A

Examining NAEP's History, Goals, Methods, And Previous Evaluations

HISTORY, GOALS, METHODS

A Historical Perspective

The origins of the National Assessment of Educational Progress can be traced back to 1867. One hundred years before the National Assessment was created, Congress authorized the establishment of the United States Office of Education and required that agency "to collect statistics and facts showing the condition and progress of education in the several states and territories and to diffuse such information respecting the organization and management of schools and school systems and otherwise promote the cause of education throughout the country" (14 Stat. 434).

For nearly a century that provision of the law received little formal attention from either the Office of Education or the Congress. This was to change in the early 1960's however, as the winds of public sentiment shifted towards an ever more insistent demand that the effects of education on students be measured, and their schools and teachers be held accountable.

The causes were many. The 1957 Sputnik launch drove the United States to intense new technological development. This in turn changed irrevocably how people earn a living in this country. Anxious parents began to look toward education to bring their children financial success, while worried citizens saw in education a way to make their country again competitive with any nation in any arena. Thus, the public scrutinized its educational system for any sign of failure to provide children with a quality education. Meanwhile, the population explosion of the post-World War II era continued to strain the capacity of schools at all levels. This led to further questions about the effectiveness of the education system.

The costs of education skyrocketed during these same years. Yet there was no evidence that the costly system was doing what anyone wanted. Statistical data such as student-teacher ratios or per-pupil expenditures could be had, but little was known about what or how much students were actually learning. There was no way to judge whether students were acquiring skills that would equip them to live and to work well in society. Nor was there a means to compare the relative effectiveness of the system from year to year.

Against this backdrop, Francis Keppel, U.S. Commissioner of Education (1962-66), set out to find a way to measure educational effectiveness and to fulfill the long-neglected legislative mandate of 1867. He asked Ralph Tyler in July 1963 to prepare a memorandum "outlining procedures by which necessary information might be periodically collected to furnish a basis for public discussion and broader understanding of educational progress and problems" (Tyler, 1971). As a result of Tyler's memorandum, the Carnegie Corporation, under the leadership of John Gardner, awarded a discretionary grant of

HISTORY, GOALS, METHODS

$12,500 to sponsor two conferences that would discuss Tyler's memorandum and "the means of ascertaining the education level attained through American Public Education."

Leading American educators and social scientists thought through the major aspects of an assessment at the first conference, convened under Tyler's direction in December 1963. Tyler believed that this first conference demonstrated that an assessment was indeed possible, but that new instruments and procedures would be needed. At the second conference, convened a month later, a consensus was reached that the time was right to attempt a national assessment.

With the sanction of these leaders, the Carnegie Corporation moved rapidly to hire John Corson of Princeton University to develop a protocol for the assessment. Corson recommended that the undertaking occur in two stages: a development stage followed by the operational one. He saw three tasks ahead during the first stage (Tyler, 1971).

Task 1. Talk with school teachers and administrators and interested laypersons to explain the assessment and to look for any likely problems.
Task 2. Contruct the instruments for the initial assessment.
Task 3. Plan how to use those instruments to gather data for the assessment.

The Carnegie Corporation confirmed its commitment to the enterprise with a $100,000 grant in June 1964. The money went to set up a committee that would coordinate these first stage tasks. Chaired by Tyler, this "Exploratory Committee on Assessing the Progress of Education (ECAPE)" pursued three charges derived from those first stage tasks identified by Tyler (Merwin and Womer, 1969).

- explore with professional educators and laypersons throughout the country ways to structure a national assessment that would secure needed information without inadvertently damaging current education practices.
- develop procedures and instruments for gathering information on the progress of education in the nation as a whole.
- suggest the make-up of a group that would put into effect the plan and procedures conceived by ECAPE, and guide later assessments of the progress of education throughout the United States.

Drawing up the details of the plan was the work of the Analysis Advisory Committee of ECAPE, whose chair was John Tukey and whose other members were Robert Abelson, Lee Cronbach, and Lyle Jones. Contracts were let for the development of different parts of the assessment. The Educational Testing Service, the American Institutes for Research, Science Research Associates, and the Psychological Corporation developed the objectives and exercises for the subject areas. The Research Triangle Institute

HISTORY, GOALS, METHODS

developed the sampling plan while the National Opinion Research Center, the Eastern Regional Institute for Education and the Southeastern Education Laboratory conducted special studies.

With its work accomplished and the trial stage complete, ECAPE became the Committee on Assessing the Progress of Education (CAPE) in 1968, signalling the start of the second, operational stage of the project. The first assessment, in Science, Citizenship and Writing, was conducted in the spring of 1969.

A final name change to the National Assessment of Educational Progress (NAEP) marked the assumption of the assessment's governance by the Education Commission of the States (ECS) in July 1969. The National Assessment program operated as an ECS project first under contract to the Office of Education until 1975, and then under contract with the National Center for Education Statistics (NCES) from 1975-78. Responsibility for the program was transferred, this time to the National Institute of Education (NIE), by the Education Amendments of 1978 (P.L. 95-561).

$59 million in federal funds has gone to the financing of the National Assessment, under the authority of the Department of Health, Education and Welfare (now the Department of Education). Other federal agencies have helped to pay for special studies. The Carnegie Corporation, the Ford Foundation and the Fund for the Advancement of Education together contributed more than $5.1 million to finance NAEP's development, planning and preparation during the years 1964-70.

Purposes And Goals

The National Assessment of Educational Progress is "a continuing national survey of the knowledge, skills, understanding and attitudes of young Americans in major learning areas usually taught in the schools. National Assessment gathers information concerning the degree to which educational goals are being met and makes this information available to the public — particularly to persons in the field of education — so that problem areas can be identified, priorities established and progress over a period of time determined." (NAEP, 1981).

Four purposes for the National Assessment program were mandated by the Education Amendments of 1978.

1. To collect and report at least once every five years data assessing the performance of students at various age or grade levels in each of the areas of Reading, Writing, and Mathematics;
2. To report periodically data on changes in the knowledge and skills of such students over time;
3. To conduct special assessments of other educational areas, as the need for addtional nationwide information arises; and

HISTORY, GOALS, METHODS

4. To provide technical assistance to state and local educational agencies as to how to use National Assessment objectives, primarily as these pertain to the basic skills of Reading, Mathematics, and Communication, and how to compare such assessments with the national profile and change data developed by the National Assessment.

The formal goals of the National Assessment are:

1. To detect the current status and report changes in the educational attainment of young Americans.
2. To report long-term trends in the educational attainments of young Americans.
3. To report assessment findings in the context of other data on educational and social conditions.
4. To make the National Assessment data base available for research on educational issues, while protecting the privacy of both state and local agencies.
5. To disseminate findings to the general public, to the federal government and to other priority audiences.
6. To advance assessment technology through an ongoing program of research and operation studies.
7. To disseminate assessment methods and materials and to assist those who wish to apply them at national, state and local levels.

The Amendment of 1978 directed the NIE to make a non-profit education organization responsible for the management of the Assessment, but delegated authority for desgin and supervision of the project to an Assessment Policy Committee (APC). NIE awarded a grant to ECS in 1979 to continue operating the National Assessment through 1983, when another competition will be held.

Methods And Procedures

Students have been measured at four age levels — nine, thirteen, seventeen and twenty-six to thirty-five — as originally recommended by ECAPE. These ages correspond, for most, to the end of the primary, junior high, and high school grades, and to the years just beyond formal schooling. Limited budgets have meant less frequent measurement of the oldest group and of those seventeen-year-olds not in school.

The National Assessments are built on an objective referenced, rather than a normed referenced, approach. Thus, the results are couched in terms of mastery of the objectives rather than of a comparison among students. The objectives and exercises which are assessed are derived through a consensus method, involving experts in the subject, educators, minority representatives and laypersons. The exercises have included multiple-choice items, attitude and self-report tasks, interviews, and performance and manipulation exercises. About 300 to 900 exercises are administered in any one assessment.

HISTORY, GOALS, METHODS

The hallmark of the NAEP model is its use of matrix sampling. No student is administered all the exercises, nor is any exercise administered to all the students. Most commonly, eight to sixteen "packages" are assembled for each in-school age-group and four to eight packages for the out-of-school groups. It takes about fifty minutes to administer a single package, each of which has about forty exercises and some background questions. Students in school are not given more than one package, while those out of school are administered as many as four. Thorough and careful steps have been taken to minimize error: administration of the exercises by a professional field staff, identical paced tapes, and scorable answer booklets.

The NAEP sampling is based on a three-stage stratified probability model. Each package is administered to anywhere from 1,000 to 2,800 individuals. Of late, the range has typically been 2,000 to 2,500 individuals. Results are usually reported for each exercise and on an aggregated composite basis as well.

Data are reported nationally and by geographic region, size and type of community, sex, race, parental education, and modal grade. About fifty percent of the exercises administered in each assessment are used again the next time the subject is assessed to permit comparisons across time.

Previous Evaluations

It goes without saying that for this assessment of the National Assessment of Educational Progress to be fair and accurate, it must include a thorough analysis of past successes, failures, and limitations of the NAEP program. During the National Assessment's brief history, it has been the object of four major evaluations. The first was conducted in 1972-73 by William Greenbaum, Michael Garet and Ellen Solomon of the Harvard Graduate School of Education. The second and third, in 1974 and 1975, were coordinated by teams assembled by NCES to evaluate the National Assessment proposals for funding the following years (1975 and 1976). The fourth and final evaluation was the work of the General Accounting Office (GAO) in 1976.

The findings of these previous evaluations are organized into five categories: goals and purposes, objectives and exercises, design, uses and users, and state and local district reporting.

Goals And Purposes

A frequently cited criticism is the alleged failure to address current educational problems. "An unholy alliance with the past ... should not be permitted to prevent NAEP from becoming a responsive instrument in imaginative efforts to solve today's

HISTORY, GOALS, METHODS

and tomorrow's educational problems" (Johnson, et. al., 1975, p. 10). Those who evaluated the project in 1974 recommended that short-term assessments be conducted. These would be based on changing national needs and use whatever assessment methods were most valid for a particular problem. The 1975 team believed that it was for the program to provide the national data pertinent to such issues, even if it meant changing the model.

The 1975 team further argued against the NAEP practice of sequestering results from policy decisions; they believed there should be a mutually beneficial relationship between the purposes of NAEP and federal policy making.

The GAO study recommended that the National Assessment officers and staff ask for ideas and suggestions from those who might use NAEP data. The study also suggested that special studies be conducted for agencies able to finance them.

Objectives And Exercises

Greenbaum and his colleagues took a close and critical look at the way decisions to assess particular subjects were reached. They pointed to Tyler's "Cardinal Principle" — of assessing what the schools were trying to accomplish — as the driving force behind those decisions to select certain subjects and objectives rather than others. The Greenbaum Report called this method "defining education as the sum of the objectives of ten subject areas," and questioned its validity.

The study also examined the objectives and exercises and found these to be neither sufficiently comprehensive nor related. "Technical problems, including large questions about the relationship of many of the exercises to the subject-matter objectives, indicate that adequate exercise development requires an effort characterized by a degree of comprehensive planning and imagination so far not evidenced" (Greenbaum, 1977, p. 71).

The way in which the National Assessment staff went about developing exercises was further criticized by the 1974 NCES report. "Exercise development at NAEP has been determined almost entirely by opinion. This is not enough. Problems of external validity, reliability, meaning and relationship remain. Results for grouped exercises must be based on a pre-determined rationale or factor structure. The possible cultural bias of exercises must be explored. An explanation of subcompetencies required to pass an exercise — based on analysis — might be useful" (Provus, et. al, 1974, p. 31).

Nor did some of the exercises even meet the standards of NAEP's designers, Greenbaum concluded. That report found that some exercises suffered from poorly coordinated or inequitably distributed objectives and sub-objectives, others from

HISTORY, GOALS, METHODS

the measurement of multiple objectives and sub-objectives, and still others from failing to measure any objectives at all.

Taking into consideration at the test development stage the aim of determining aggregate scores on clusters of exercises was recommended by the NCES and GAO studies. They also suggested that the clustering be tested to see whether it was meaningful and comprehensible to the curricular objectives and practices of most schools.

Design

There has been high praise for the technological advances and contributions to the measurement field made as a result of the National Assessments, especially in the areas of sampling, the development of objective referenced assessment instruments, standardized administrations, improved quality control over data collection, innovative statistical analysis techniques, and new approaches to exercise writing. But criticisms were also raised of NAEP's design, relationship to research, failure to use background variables, and the limitations of matrix sampling.

It was frequently recommended that researchers be involved more extensively in all NAEP phases. The 1975 team recommended the inclusion of both technical and social policy studies in NAEP research, and concluded that policy relevant research could be a "highly promising approach to increasing both the usefulness and visibility of NAEP" (Johnson, et. al. 1975, p. 18).

Similar concern was heard on the wisdom of asking different geographic, demographic, and other background questions, and disaggregating the data more finely. Greenbaum and colleagues even suggested that if there were adopted a design that examined the performance of individuals rather than groups, it might advance learning theory, delineate specific research questions, or suggest new curriculum approaches.

Flexibility was urged. "To the extent that it is, or becomes, unthinkable to abort a cycle once development has commenced, to reduce the number of in-school age levels assessed, to lengthen the time between assessments, to assess on some basis other than subject areas, or to report on some basis other than exercises — to that extent the mainline function (the usual model) controls NAEP rather than vice versa" (Johnson, et. al. 1975, p. 14).

Uses And Users

Everyone pretty much agreed that the results of National Assessment have never been very useful in a practical sense. The GAO study recommended some steps to change this:

HISTORY, GOALS, METHODS

1. identify the information and other needs of decision makers;
2. determine the feasibility and cost effectiveness of alternative approaches to satisfy those needs;
3. decide on the assessment approach to be used;
4. establish continuous dialogues to determine data needs and how NAEP could best meet them;
5. interpret data rather than relying on outside interpretation;
6. establish performance standards;
7. improve communication with NIE to facilitate research and
8. improve dissemination of project results.

Another major concern has been the program's failure to interpret the results of the data gathered. "NAEP should realize that it is already engaged in interpretation (by virtue of the way in which it currently summarizes and reports data) and rather than stoutly deny such interpretations, it could acknowledge the practice and become more sophisticated about how to do it" (Provus, et. al., 1974, p. 29).

Lastly, GAO criticized the lack of performance standards for the NAEP program. "A factor contributing to the lack of interpretation of National Assessment data is the lack of standards against which test data can be compared to judge performance...The National Assessment contends that no one knows for sure what a reasonable percentage of success should be, partly because concrete achievement data has (sic) never been available...In our opinion, unless meaningful performance comparisons can be made, state, localities, and other data users are not as likely to find the National Assessment to be useful." (GAO, 1976, p. 28).

State And Local District Reporting

At no time has the Assessment reported data for individual states or school districts. All kinds of factors have prevented this: confidentiality, the proportionately large sample size required to measure relatively small populations, and uncertainty as to what the appropriate federal government role in education is. The 1975 NCES report suggested that while insufficient collection of state and local data might have been understandable in 1965, it was no longer in 1975.

HISTORY, GOALS, METHODS

In the mid-sixties, the emphasis was on national curriculum development projects to improve student achievement. A decade later public sentiment was shifting toward greater control and decision making by the states and local districts. At about the same time emphasis shifted away from subject-oriented curriculum efforts towards a vision of the classroom as a place where all society's problems might be solved. These policy changes — the 1974 NCES report observes would have been best served if NAEP officers had adjusted the policy so as to provide state and local data. "We regard this as most unfortunate (the lack of state and local level data) since we believe that improvement of education must be done at the local level and within particular classrooms and schools." (Provus, et. al., 1974, p. 4).

Summary

All of the evaluations were fundamentally consistent with the conclusion reached by the 1974 evaluation team:

We believe that the survival and strengthening of the project is vital to the continuing effort to improve the quality of education of American school children ...But its potential is still to be realized. National Assessment has not been widely used as a source of benchmarks on the effectiveness of American education by state and local educators, nor as an effective tool in the education policy making process of officials of the legislative and executive branches of government at either the state or national level, nor by research scholars who affect the content or curricula and teacher training methods, nor by the producers of textbooks, nor by large numbers of school administrators and classroom teachers...The basic need is for NAEP to adopt a more dynamic and more flexible approach both to its purposes and to its procedures for implementing them.

APPENDIX B

We
Owe
Thanks
To . . .

WE OWE THANKS TO...

We are indebted to the following group of distinguished people who provided us with invaluable information and counsel through personal or telephone interviews, conversations, responses to our queries, or critiques of working documents. Our debt to them is undiminished by the understandable fact that their views are not always those expressed in the report.

Stanley Ahmann, Iowa State University
Peter Airasian, Boston College
Gordon Ambach, New York Department of Education
Robert Andringa, Education Commission of the States
Alexander Astin, University of California at Los Angeles
Richard Atkinson, University of California at San Diego
Eva Baker, University of California at Los Angeles
Joan Baratz, Education Policy Research Institute
Ernest Bauer, Oakland School District, Michigan
David Bayless, Research Triangle Institute

Honorable Terrell Bell, United States Secretary of Education
Jon Bentz, NAEP Assessment Policy Committee
Stanley Bernknopf, Georgia Department of Education
Marietta Blackburn, NAEP Assessment Policy Committee
Benjamin Bloom, University of Chicago
Clarence Blount, NAEP Assessment Policy Committee
Ernest Boyer, Carnegie Foundation for the Advancement of Teaching
George Brain, Washington State University
Alfred Brennan, Scott, Foresman & Company
Mervin Brennan, Illinois Department of Education
D. Allan Bromley, American Association for the Advancement of Science
Rexford Brown, National Assessment of Educational Progress
Edward Bryant, WESTAT, Inc.
Fred Burke, New Jersey Department of Education
George Burkett, CTB/McGraw-Hill
Gilbert Bursley, NAEP Assessment Policy Committee
John Carroll, University of North Carolina
Gordon Cawalti, Association for Supervision and Curriculum Development
Jeanne Chall, Harvard University
William Clemans, American Institutes for Research
William Coffman, University of Iowa

Robert Coldiron, Pennsylvania Department of Education
Beverly Cole, National Association for the Advancement of Colored People
William Connett, Montana Department of Education
Frank Corrigan, National Center for Education Statistics
Lee Cronbach, Stanford University
Joseph Cronin, Massachusetts Student Financial Aid
Keith Cruse, Texas Department of Education
Oluf Davidson, American College Testing Program
Nancy Dearman, National Center for Education Statistics
Thomas Donlon, Educational Testing Service
David Donovan, Association for Measurement and Evaluation in Guidance
Richard Duran, Educational Testing Service
Marie Eldridge, National Center for Education Statistics
William Eller, International Reading Association
Marian Epstein, Educational Testing Service
Marion Faldet, The Spencer Foundation
Roger Farr, Indiana University
A.L. Finkner, Research Triangle Institute

WE OWE THANKS TO...

Thomas Fisher, Florida Department of Education
Thomas Fitzgibbon, The Psychological Corporation
John Flanagan, American Institutes for Research
Jerry Floyd, National School Boards Association
Roy Forbes, National Assessment of Educational Progress
Garlie Forehand, Educational Testing Service
Pascal Forgione, Connecticut Department of Education
Fred Forster, Portland School District, Oregon
Gloria Frazier, National Assessment of Educational Progress
Mary Futrell, NAEP Assessment Policy Committee
June Gabler, NAEP Assessment Policy Committee
James Gates, National Council of Teachers of Mathematics
Gene Geisert, NAEP Assessment Policy Committee
Richard Gereau, Staff to Senator Claiborne Pell
Carol Gibson, National Urban League
Robert Glaser, University of Pittsburgh
John Goodlad, University of California at Los Angeles
Patricia Graham, Harvard Univesity
Thomas Green, Syracuse University
John Guth, SRA/IBM
Monsignor James Habiger, NAEP Assessment Policy Committee

Constance Hadley, National Assessment of Educational Progress
Walter Haney, Huron Institute
George Hanford, The College Board
James Hazlett, F.T. Jones & Co., Kansas City, Missouri
Madeline Hemmings, The Chamber of Commerce
John Hemphill, Far West Laboratory for Educational Research and Development
Martha Highsmith, Rhode Island Department of Education
John Hopkins, Research for Better Schools
Jerry Horn, Association for the Education of Teachers in Science
Torsten Husén, University of Stockholm
Thomas Innes, University of Tennessee
Stephen Ivens, The College Board
Richard Jaeger, University of North Carolina at Greensboro
Thomas James, The Spencer Foundation
Robert Janus, Riverside Publishing Company
Martin Jensen, Staff to Committee on Labor and Human Resources
George Johnson, The Riverside Publishing Company
Theodore Kaltsounis, National Council for Social Studies
Eugenia Kemble, American Federation of Teachers
Francis Keppel, Harvard University
Keith Kershner, Research for Better Schools

Barbara Klein, NAEP Assessment Policy Committee
Donna Knight, NAEP Assessment Policy Committee
Nancy Kober, Staff to Rep. Carl Perkins, Subcommittee on Elementary, Secondary, and Vocational Education
Gerald Koch, NAEP Assessment Policy Committee
Roger Lennon, Harcourt, Brace, Jovanovich, Inc.
Frederic Lord, Educational Testing Service
Winsor Lott, New York Department of Education
Ruth Love, Chicago School District
C. LaVor Lyn, Dallas School District
Frank Macchiarola, New York City School District
George Madaus, Boston College
Wayne Martin, National Assessment of Educational Progress
Marjorie Martus, The Ford Foundation
Horace Maxcy, Maine Department of Education
Wilbert McKeachie, University of Michigan
Floretta McKenzie, District of Columbia School District
James Mecklenberger, National School Boards Association
Jack Merwin, Univesity of Minnesota
Jon Miller, Northern Illinois University

WE OWE THANKS TO...

Juliet Miller, Association for Measurement and Evaluation in Guidance
Jason Millman, Cornell University
Martin Milrod, National Institute of Education
Anita Mitchell, Association for Measurement and Evaluation in Guidance
Thomas Montebell, West Virginia Department of Education
Frederic Mosher, The Carnegie Corporation
Ina Mullis, National Assessment of Educational Progress
Carol Norman, National Education Association
Trudi Odbert, NAEP Assessment Policy Committee
Susan Oldefendt, National Assessment of Educational Progress
Michael Olivas, LULAC National Education Service Centers
Thomas Owens, Association for Experiential Education
Virginia Plisco, National Center for Education Statistics
Arthur Pontarelli, Rhode Island Department of Education
Robert Rath, Northwest Regional Educational Laboratory
Daniel Resnick, Carnegie-Mellon University
Edward Rincon, National Congress of La Raza
Alan Robertson, Association for Measurement and Evaluation in Guidance
Edward Roeber, Michigan Department of Education
Joseph Romero, NAEP Assessment Policy Committee
Phillip Runkel, Michigan Department of Education
Charlotte Ryan, NAEP Assessment Policy Committee
Jeffrey Schiller, National Institute of Education
Jack Schmidt, National Assessment of Educational Progress
Richard Schutz, Southwest Regional Laboratory for Educational Research and Development
Dunlap Scott, National Assessment of Educational Progress
Donald Searls, National Assessment of Educational Progress
Penny Sebring, Northwestern University
Thomas Shannon, National School Boards Association
Roger Sharpe, Harvard University
Lee Shulman, Michigan State University
Lynn Simons, NAEP Assessment Policy Committee
Theodore Sizer, Cambridge, Massachusetts
John Slaughter, National Science Foundation
Robert Smith, Council for American Private Education
Frank Snyder, CTB/McGraw-Hill
Robert Solomon, Educational Testing Service
Robert Stake, University of Illinois at Urbana-Champaign
Floretta Stevens, Los Angeles School District
Donald Stewart, Fountain Valley, California
Robert Stonehill, United States Department of Education
A.A. Strassenberg, American Association for Physics Teachers
Dorothy Strickland, Teachers College, Columbia University
Phillip Swain, NAEP Assessment Policy Committee
Walter Tice, NAEP Assessment Policy Committee
David Tiedeman, National Institute for Career Education
P. Michael Timpane, Teachers College, Columbia University
Ralph Turlington, Florida Department of Education
William Turnbull, Educational Testing Service
Ralph Tyler, SRA/IBM
Janet Wall, Department of Defense Dependent Schools
Barbara Ward, National Assessment of Educational Progress
David Wiley, Northwestern University
Frank Womer, University of Michigan
David Wright, National Assessment of Educational Progress

REFERENCES

REFERENCES

Commissioned By Wirtz And Lapointe

Bryant, E.C. Measurement of changes in educational achievement for the nation.

Gibson, C. What is the relationship of a national assessment to the goal of improving educational opportunity for all students?

Graham, P.A. Changes in the educational environment of the United States: 1964-1981.

Lord, F.M. Advantages/disadvantages of the current National Assessment design.

Timpane, P.M. What is the feasibility of establishing a national consensus on what students should learn?

General

Ahmann, J.S. & Larson, R. Using item banking procedures for assessing changes in levels of achievement. In de Gruijter & van der Kamp (Eds.), *Advances In Psychological and Educational Measurement*. New York: John Wiley & Sons, 1976

Andrews, F.M. & Withey, S.B. *Social Indicators of Well Being*. New York: Plenum Press, 1976.

Baker, E. *Toward local control and national accountability in federal program evaluation*. Paper commissioned for the National Institute of Education's working conference on Planning the Evaluation Process for the Follow Through Program, Texas, February 1981.

Baker, E. & Quellmalz, E.S. *Educational Testing and Evaluation*. Beverly Hills, California: Saga Publications, 1980.

Bauer, R.A. (Ed.) *Social Indicators*. Cambridge, Massachusetts: The M.I.T. Press, 1966.

Bayless, D.L. & Nix, C.W. *Competency testing: setting educational performance standards for the group*. Paper presented at the ninth Annual Conference on Large-Scale Assessment, Denver, June 1979.

Black, P. & Marjoram, T. *National and state assessment in the U.S.A.* London: Department of Education and Science, March 1979.

Boruch, R.F. *Notes on the National Assessment of Educational Progress* Personal Communication; August 14, 1980.

Brain, G.B. *National Assessment — evaluation and accountability*. Paper presented at the National Association of State Boards of Education meeting, October 1970.

Brain, G.B. Some values of assessment. *Compact,* February 1972, pp. 5-6.

Burstall, C. & Kay, B. *Assessment — the American experience*. London: Department of Education and Science, January 1978.

Burton, N. *Assessment as exploratory research: a theoretical overview*. Paper presented at the annual meeting of the American Educational Research Association, Toronto, March 1978.

Burton, N.W. Societal standards. *Journal of Educational Measurement*, 1978, *15*, 263-261.

Chapin, J. Using the NAEP test exercises. *Social Education*, 1974, *38*, 412-414.

REFERENCES

Chromy, J.R. & Horvitz, D.G. The use of monetary incentives in National Assessment household surveys. *Journal of the American Statistical Association*, 1978, *73*, 473-478.

Comptroller General of the United States. *The National Assessment of Educational Progress: its results need to be more useful*. Washington, D.C.: General Accounting Office, July 1976.

Cronbach, L.J. Five decades of public controversy over mental testing. *American Psychologist, 1975, 30,* 1-14.

Cronbach, L.J., et al. *Toward Reform of Program Evaluation*. San Francisco, California: Jossey-Bass, Inc., 1980.

Cross, C.T., Gwaltney, M.K., Lee, J.B. & Weiss, S. *Improving the usefulness of the National Assessment of Educational Progress for federal policy-makers and national associations.* (Contract No. 02-80-18662). Cambridge, Massachusetts: ABT Associates, June 1980.

Davis, J.A. & Collins, E. *Objectives, design and history of the National Longitudinal Study.* Paper presented at the annual meeting of the American Educational Research Association, Washington, D.C., April 1975.

Dyer, H. *Parents can understand testing.* Columbia, Maryland: The National Committee for Citizens in Education, 1980.

Ebel, R. *Essentials of Educational Measurement*. Englewood Cliffs, New Jersey: Prentice-Hall, 1972.

Fair, J. What is National Assessment and what does it say to us? *Social Education,* 1974, *38,* 398-414.

Farr, R. & Fay, L. *Then and Now: Reading achievement in Indiana (1944-45 and 1976).* Bloomington, Indiana: School of Education, Indiana University, 1978.

Fienberg, S.E. & Mason, W.M. Identification and estimation of age-period-cohort models in the analysis of discrete archival data. In Schuessler (Ed.) *Sociological Methodology,* San Francisco: Jossey-Bass, Inc., 1979.

Finley, C.J. National Assessment reports and implications for school districts. *The National Elementary Principal,* 1971, *50,* 25-32.

Finley, C.J. Not just another standardized test. *Compact,* February 1972, pp. 9-12.

Forbes, R.H. *National Assessment: policy decision information.* Paper presented at the American Educational Research Association annual meeting, Boston, April 1980.

Foxman, D.C., Cresswell, M.J., Ward, M., Badger, M.E., Tuson, J.A., Bloomfield, B.A. *Mathmatical development: primary survey report No.1.* London: Assessment of Performance Unit, Department of Education and Science, 1980.

Glaser, R. The future of testing: A research agenda for cognitive psychology and psychometrics. *American Psychologist.* 1981, *36,* 923-936.

Gorth, W.P. & Perkins, M.R. *A study of minimum competency testing programs: comprehensive report.* Amherst, Massachusetts: National Evaluation Systems, December 1979.

Greenbaum, W., Garet, M. & Solomon, E. *Measuring Educational Progress.* New York: McGraw-Hill Book Company, 1977.

Hambleton, R.K. Latent trait models and their applications. *New Directions for Testing and Measurement,* 1979, *4,* 13-32.

Hand, H.C. The camel's nose. *Phi Delta Kappan,* 1965, *47,* 8-13.

Harnischfeger, A., Huckins, L. & Wiley, D. *The National Assessment of Educational Progress model: A tool for achievement based Title I fund allocations.* Illinois: ML-Group for Policy Studies in Education, CEMREL, Inc., 1977.

Harnischfeger, A. & Wiley, D. *Criterion referenced tests: what they are and what they could be.* Illinois: ML-Group for Policy Studies in Education, CEMREL, Inc., 1977.

REFERENCES

Hazlett, J.A. *A history of the National Assessment of Educational Progress, 1963-1973.* Unpublished doctoral dissertation, University of Kansas, 1973.

Higgins, M.J. & Merwin, J.S. Assessing the progress of education. A second report. *Phi Delta Kappan* 1967, *48*, 378-380.

Houts, P.L. (ed.) *The Myth of Measurability.* New York: Hart Publishing Company, 1977.

Hunkins, F.P. Exercises to assess social studies and citizenship: How good are they? *Social Education*, 1974, *38*, 415-421.

Jaeger, R.J. & Tittle, C. *Minimum Competency Achievement Testing,* Berkeley, California: McCutcheon Publishing Company, 1980.

Johnson, G.H. Making the data work. *Compact,* April 1972, pp. 39-30.

Johnson, M. et. al., *An evaluation of the National Assessment of Educational Progress.* Washington, D.C.: National Center for Education Statistics, 1975.

Justus H. Findings in the National Assessment science survey. *American Education*, 1972, *8*, 7-10.

Keppel, F. *Basic public policy on student achievement: a new agenda item.* Paper presented to the American Association for Higher Education, March 1980.

Kerins, C.T. & Fyans, Jr., L.J. *The use, relevance and appropriateness of tests for education,* Personnal Communication, August 1980.

Lazarsfeld, P.F. Notes on the history of quantification in sociology — trends, sources and problems. In Woolf (Ed.), *Quantification: A History of the Meaning of Measurement in the Natural and Social Sciences.* Indianapolis, Indiana: Bobbs-Merrill, 1961.

Martin, W.H. National Assessment of Educational Progress. *New Directions for Testing and Measurement.* 1979, *2*, 44-65.

Mehrens, W.A. & Ebel, R.L. Some comments on criterion-referenced and norm-referenced achievement tests. *Measurement in Education.* 1979, *10*. (No. 1.)

Merwin, J.C. & Womer, F.B. Evaluation in assessing the progress of education to provide bases of public understanding and public policy. In National Society for the Study Education: *Educational Evaluation: New Roles, New Means 68th Yearbook, Part II.* Chicago: University of Chicago Press, 1969.

Milrod, M. *Report of the NIE grant application review panel.* Paper presented at the Assessment Policy Committee meeting, February 1980.

Mushkin, J.S. *National Assessment and social indicators.* ([OE] 73-1111). Washington, D.C.: U.S. Department of Health, Education and Welfare, 1973.

National Assessment of Educational Progress (NAEP) Grant Competition, Washington, D.C.: National Institute of Education, 1979.

Norris, E.L. National Assessment: an information gathering and information disseminating project. *Education,* 1971, *91*, 286-291.

Organization for Economic Co-operation and Development. *Science, Growth and Society — A New Perspective.* Report on the Secretary-General's Ad Hoc Group on New Concepts of Science Policy. Paris: OECD, 1971.

Pipho, C. *Update VIII: Minimum Competency Testing.* Denver, Colorado: Education Commission of the States, July 1979.

Provus, M. et al. *An evaluation of the National Assessment of Educational Progress by the Site Team established by the National Center for Education Statistics.* Washington, D.C.: Evaluation Research Center, National Center for Education Statistics, June 1974.

Public Law 95-561, (STAT. 2143) 95th Congress, November 1, 1978.

REFERENCES

Resnick, D.P. Minimum competency testing historically considered. *Review of Research in Education.* 1980, *8,* 3-29.

Resnick, D.P. Educational Policy and the applied historian. *Journal of Social History.* 1981. *14.*

Resnick, D.P. & Resnick, L.B. The nature of literacy: An historical exploration. *Harvard Educational Review,* 1977, *47,* 370-385.

Schwartz, J.L. & Garet, M.S. (Eds.). *Assessment for Accountability.* The report of a study panel to the Ford Foundation and the National Institute of Education, January 1981.

Schwartz, J.L. & Garet, M.S. (Eds.) *Assessment in the Service of Instruction.* The report of a study panel to the Ford Foundation and the National Institute of Education, January 1981.

Sebring, P.A. and Boruch, R.F., *The Use, Abuse and Nonuse of NAEP: Case Studies on the Data, Methods, and Materials.* (Report A-137-1, draft), Evanston, Illinois: Northwestern University, November 1981.

Shafer, R.E. *National Assessment: background and projections — 1975.* Paper presented at the annual meeting of the Conference on English Education. Colorado Springs, Colorado, March 1975.

Sirotnik, K.A. Introduction to matrix sampling for the practitioner. In J.W. Popham (Ed.): *Evaluation in Education — Current Applications.* Berkeley, California: McCutcheon Publishing Company, 1975.

Stake, Robert E. National Assessment *Proceedings of the 1970 Invitational Conference on Testing Problems — The Promise and Perils of Educational Information Systems.* Princeton, New Jersey: Educational Testing Service, 1971.

Stetz, F.P. & Beck, M.D. Attitudes toward standardized tests: Students, teachers and measurement specialists. *Measurement in Education,* 1981, *12,* (No. 1).

Taylor, B.L., Implications of the National Assessment model for curriculum development and accountability. *Social Education* 1974, *38,* 404-408.

Taylor, B.L. *The National Assessment Model.* Paper presented at the National Council of the Social Studies, San Francisco, 1973.

Tierney R. J. & Lapp, D. *National Assessment of Educational Progress in Reading.* Newark, Delaware: International Reading Association, 1970.

Tyler, R.W. Assessing the Progress of Education. *Phi Delta Kappan,* 1965, *47,* 13-16.

Tyler, R.W. National Assessment: a history and sociology. In J.W. Guthrie & E. Wynne (Eds.): *New Models for Education.* Englewood Cliffs, New Jersey: Prentice Hall, Inc., 1971.

Tyler, R.W. Why evaluate education? *Compact,* February 1972, pp. 3-4.

Tyler, R.W. *Role of assessment in improving educational programs.* Paper presented at the annual meeting of the American Educational Research Association, Boston, April 1980.

Tyler, R.W. & White, S.H. *Testing, Teaching and Learning.* Washington, D.C.: U.S. Department of Health, Education and Welfare, October, 1979.

Wiley, D.E. *Policy-responsive evaluation.* Paper presented at the Winter Invitational Conference on Measurement and Methodology, Center for the Study of Evaluation, UCLA, Los Angeles. January 1978 (revised April 1978).

Wiley, D.E. *Potential of state and national assessment information for improving policy development.* Paper presented at the annual meeting of the American Educational Research Association, Boston, April 1980.

Wiley, D.C. Improving policy development. *New Directions for Testing & Measurement,* 1981, *10,* 49-64.

Williams, B.I. & Gilliard, J. One more time: NAEP and blacks. *Social Evaluation.* 1974, *38,* 422-424.

Womer, F. *Developing a large scale assessment program.* Denver, Colorado: Cooperative Accountability Program, Colorado Department of Education, 1974.

REFERENCES

Womer, F. National Assessment says. *Measurement in Education.* 1970, *2,* (No. 1).

Womer, F. *What is National Assessment?* Denver, Colorado: Education Commission of the States, 1968.

Womer, F. & Mastie, M. How will National Assessment change American education? *Phi Delta Kappan,* 1971, *53,* 118-120.

National Assessment Of Educational Progress Documents

Brown, R. *Contributions of the national assessment to understanding the problems of literacy and equity* (12-IP-52).

Brown, R. *Literacy in America: A synopsis of National Assessment findings* (SY-FL-50).

Changes in mathematical achievement, 1973-80.

Changes in political knowledge and attitudes, 1969-76.

Cobb, H. *Data base management needs of National Assessment and ways to meet those needs* (12-IP-58).

Design and Development Document.

Education for citizenship.

Functional literacy: basic reading performance.

Has Title I improved education for disadvantaged students? Evidence from three National Assessments of reading.

Holmes, B.J. *Bias: psychometric and social implications for the National Assessment of Educational Progress* (12-IP-54).

Math fundamentals: selected results from the first National Assessment of mathematics.

Mathematical applications.

Mathematical knowledge and skills.

Mathematical understanding.

National Assessment of Educational Progress, Annual Report, 1980.

National Assessment of Educational Progress Staff. *Issues in the Analysis and Analysis of Change of National Assessment data:* (No. 12-IP-57).

National Assessment of science, 1979-73.

Q & A about the National Assessment of Educational Progress.

Reading and mathematics achievement in public and private schools: Is there a difference?

Reading in America: A perspective on two assessments.

Reading Objectives.

Reading, writing and thinking.

Recipes, wrappers, reasoning and rate. A digest of the first reading assessment.

Science achievement in the schools.

Scott, Jr., Dunlap. *Access to Schools and Nonstudents.* (12-IP-56).

Three National Assessments of reading: Changes in performance, 1970-80.

Ward, B. *Major informational needs of National Assessment audiences and ways to enhance the assessment's utility in meeting those needs (12-IP-51).*

Ward, B. *The National Assessment approach to objectives and exercise development* (12-I-55).

Writing achievement, 1969-79 (Vols. I, II, and III).